CIRIA C573 London, 2002

A guide to ground treatment

J M Mitchell

F M Jardine

This report was largely completed before 1993 but has been reviewed and updated in the light of comments made in 1999 and 2000 by the steering group for CIRIA Research Project 604, "Treated ground – engineering properties and durability" (issued as CIRIA publication C572, 2002).

CIRIA *sharing knowledge* ■ *building best practice*

6 Storey's Gate, Westminster, London SW1P 3AU
TELEPHONE 020 7222 8891 FAX 020 7222 1708
EMAIL enquiries@ciria.org.uk
WEBSITE www.ciria.org.uk

Summary

An introduction to ground treatment methods, this report explains the available techniques and their variations by systematic descriptions of the physical principles, the equipment, the methods and the effectiveness that can be expected for each. General guidance is given on the matters to be considered when ground improvement is being considered as an option. Particular attention is given to the responsibility for design and the roles of those involved in the design process and in control of the treatment. Guidance is given on selection of appropriate techniques by reference, where possible, to comparative studies and to case histories. Individual techniques are described separately, with key references on their design and use. The techniques are grouped as those that achieve improvement by vibration, adding load, structural reinforcement, structural fill, admixtures, grouting, thermal methods and vegetation.

A guide to ground treatment

Mitchell, J M and Jardine, F M

Construction Industry Research and Information Association

CIRIA publication C573   © CIRIA 2002   ISBN 0 86017 573 1

Keywords		
Ground improvement, piling		

Reader interest	Classification	
Design, specification, construction, managers, clients and supervising engineers involved in civil and geotechnical works	Availability	Unrestricted
	Content	Subject area review
	Status	Committee-guided
	User	Civil, geotechnical and structural engineers, engineering geologists

Published by CIRIA, 6 Storey's Gate, Westminster, London SW1P 3AU.

Foreword

This report presents the results of a research project carried out for CIRIA by the late J M Mitchell of Ove Arup and Partners. His untimely death prevented John Mitchell from completing both this report and what was intended as a more detailed accompanying report on deep compaction techniques. John had assembled much more material than is given in this report and, prior to preparing the final document, had reached the stage of selecting what should be included and what left out. In this, he and CIRIA's research manager, F M Jardine, had been working closely together and had agreed the general structure and much of the content and to share authorship. Since then, with the help of John's colleagues at Arup Geotechnics, the second author took over completion of the report. It is hoped that it does justice to John's work and to his intentions for it.

This report was largely completed before 1993 but has been reviewed and updated in the light of comments made in 1999 and 2000 by the steering group for CIRIA Project 604 "Treated ground: engineering properties and durability".

Following CIRIA's usual practice, this project was guided by a steering group, which during the course of the project comprised:

Mr C Garrett (chairman)	then of Kent County Council
Dr A L Bell	Keller Ground Engineering Ltd
Dr J A Charles	Building Research Establishment
Dr D A Greenwood	then of Cementation Piling & Foundations Ltd
Dr R A Jewell	then of GeoSyntec Consultants
Dr L M Lake	then of Mott MacDonald Group
Mr M P Moseley	Keller Ground Engineering Ltd
Mr N A Trenter	then of Sir William Halcrow & Partners.

CIRIA's research manager for the project was Mr F M Jardine.

Although the report does not include post-1993 modifications of techniques or many recent case histories, it has been reviewed by the CIRIA steering group for Research Project 604, who recommended its publication and provided comment. Some additional notes are given in italics at the start of sections to draw attention to recent changes not fully addressed in the report.

Note

Government reorganisation has meant that DETR responsibilities have been moved variously to the Department of Trade and Industry (DTI), the Department for the Environment, Food and Rural Affairs (DEFRA), and the Department for Transport, Local Government and the Regions (DTLR). References made to the DETR in this publication should be read in this context.

For clarification, readers should contact the Department of Trade and Industry.

ACKNOWLEDGEMENTS

The project was funded by Construction Directorate of the Department of the Environment, Transport and the Regions, by CIRIA and by CIRIA's Core Programme.

CIRIA is grateful for the help and advice given to the project by many firms and individuals and particularly those who shared their knowledge with John Mitchell in many discussions about ground treatment techniques. He would have wished to acknowledge these individuals by name, but as we cannot be sure of including all, we hope they will understand the acknowledgement not being specific.

The help of members of the RP604 steering group who checked and provided additional text is also gratefully acknowledged, particularly Dr D H Beasley of Halcrow Group Limited, T J P Chapman of Ove Arup and Partners, S Everton of Gibb Limited, Mr W M Kilkenny of WS Atkins Consultants, Mr C Raison of Chris Raison Associates, Dr J M Reid of TRL, and T Schofield of Stent Foundations Limited.

CIRIA and the authors gratefully acknowledge the support of these funding organisations and the technical help and advice provided by the members of the steering group. Contributions do not imply that individual funders necessarily endorse all views expressed in published outputs.

Acknowledgement is made to the following organisations for permission to use their photographs:

Figure 3.2	Ove Arup & Partners, and air photo Hunting Surveys Ltd
Figure 4.9	GKN Keller, now Keller Foundations Ltd
Figure 4.11	Cementation Piling and Foundations Ltd, now Kvaerner Foundations Limited
Figure 4.13	Dr H J Walbancke, Binnie, Black and Veatch
Figure 4.14	Dr H J Walbancke, Binnie, Black and Veatch
Figure 4.15	Ove Arup & Partners
Figure 4.18	BRE
Figure 6.10	F M Jardine.

Contents

APPENDICES

Tables

Figures

1 Introduction

CIRIA is publishing a series of reports on particular forms of ground improvement processes, ie dewatering (Preene *et al*, 2000), vertical drains (Holtz *et al*, 1991), geotextiles for soil reinforcement (Jewell, 1996), and geotechnical grouting (Rawlings *et al*, 2000). There are many other techniques for ground improvement, less widely used but capable of providing economic technical solutions for the many types of poor ground. This report is an introduction to ground improvement methods; it sets out the available techniques and their variations by systematic descriptions of the principles, the equipment, the methods and the effectiveness that can be expected for each. The report, therefore, is a general guide to the available techniques by which ground can be improved and to the situations for which particular techniques are appropriate.

Since the start of the research project on which this report is based, several important texts have been published on this and associated subjects, eg *Ground Improvement* (ed M P Moseley, 1993), *Soil improvement techniques and their evolution* (Van Impe, 1989), and *Building on fill: geotechnical aspects* (Charles, 1993). Under the auspices of the International Society for Soil Mechanics and Geotechnical Engineering, Thomas Telford Ltd now publishes a quarterly journal entitled *Ground Improvement*. In addition, there have been several speciality conferences, such as the three on ground improvement systems, again sponsored by the International Society through its technical committee on this subject, TC-17. The committee's website (http://tc17.poly.edu/ikd.htm) includes descriptions of improvement techniques and a database.

In 1999, CIRIA started a new research project to provide guidance about the engineering properties and long-term performance of treated ground. That report, CIRIA publication C572 *Treated ground: engineering properties and performance* (Charles and Watts, 2002), complements and amplifies the content of this one, without having to repeat the explanations and descriptions of the techniques given here.

1.1 THE OPTION OF IMPROVING THE GROUND

Virtually all engineering construction involves the ground. In poor ground conditions there are five options:

- to bypass the poor ground, by moving to a new site, or using deep foundations to stronger ground
- to remove the poor ground, replacing it with better material
- to design the structure to allow for the behaviour of the poor ground under load
- to treat the poor ground to improve its properties (ie ground improvement)
- to abandon the project (the promoter's decision).

The fourth option, of ground treatment, gives considerable scope to engineers for finding a viable solution to the problems of poor ground. A wide range of treatments is available, techniques can be selected and combined to cope with different aspects of the poor ground, and there is increasing confidence both in what can be achieved by well-executed treatment and in its proper integration into the overall scheme for the construction. All these points are evidence of how valuable is this option.

The objective of treatment is improvement. When ground treatment is being considered as an option, it is important that all who will be involved in it should recognise not only what can reasonably be achieved by a particular technique, but also the extent of their responsibility if it is chosen.

For this reason, therefore, the first three sections of this report address issues general to all ground improvement schemes. The issues are of responsibility for design, of site investigation, of assessing and minimising the risks inherent in the chosen ground treatment scheme, and of matters relating to placing and completing a treatment contract. The recommendations given on these issues are for good practice; as such, they are intended for the benefit of promoters, project design professionals and specialist ground engineering contractors.

It is the variability of the ground – which is not only the engineering material itself, but also the medium in which the work takes place – that makes ground engineering an iterative process of discovery and innovation. When the objective of the work is to improve the ground in specific ways, success – and demonstrating it – depends upon first knowing what the existing ground is like and then, during the work and later, upon evaluating the effect of the treatment technique. Control testing and assessment of the effectiveness of the treatment are thus essential components in a scheme of ground improvement. There may also be reason to monitor the long-term performance of the treated ground and of structures built on it.

Aspects that have to be considered in selecting the appropriate ground improvement technique are discussed in general terms (Section 3.3), in the descriptions of individual techniques given in Sections 4 to 11 inclusive, and in relation to the combinations of different methods (Section 12) and to the results of comparative studies published by others (Appendix B).

In Sections 4 to 11, the techniques of ground treatment are explained under the following headings:

- definition, of the terms used for the technique
- principle, of the mechanics by which the treatment takes effect
- description, of the methods used
- applications, for which the technique is used
- limitations, that could affect the suitability of the technique
- design, by pointers to sources of design guidance
- control, of the treatment works by tests and measurements.

These explanations are necessarily introductory, but references are provided that give further and more detailed information. For techniques that are essentially standard civil engineering operations, such as earthworks, there is no text section on controls.

It is important to recognise that the limitations are not necessarily sufficient reason for a technique to be excluded from consideration. Modification to the construction method or design may be possible – the point being that these possible limitations should be appreciated at the feasibility stage of a project. It is the nature of many ground treatment techniques that their capability is continually being extended, overcoming what were previously seen as limitations. Moreover, different techniques can be combined to cope with a greater range of situations than one method on its own.

1.2 GROUND IMPROVEMENT

The term "ground improvement" is open to different interpretations. First, it is an intention or objective, not the process of achieving it, although the term is often used in that sense. Second, improvement is a relative condition as to which aspect and to what degree there is improvement.

From the outset of this CIRIA project there was debate in the steering group guiding the research and in the minds of the authors about what the report should be included in the report. Should the coverage be restricted to those treatment techniques that alter the state or nature of the *in-situ* ground materials, in the way, for example, that deep compaction by increasing their density changes the state of granular soils, and permeation of grouting changes the nature of ground by sealing its pores? If changing the ground's nature at the micro scale is easily recognisable as resulting in an improvement of its condition, what if the treatment is of a large mass of ground by the inclusion of discrete reinforcing elements that, in themselves, do not alter either the state or nature of the ground strata? The best example of this is piling, which would not usually be thought of as ground treatment. A third question relates to the improvement of fill materials: for example, are reinforced soil embankments, reinforced foundation soil or track beds examples of ground improvement? Even less clear is the use of lightweight fills instead of increasing the stiffness and strength of the underlying weak ground. This technique has been included if only for the reason that its use is often in combination with foundation improvement.

For the purposes of the coverage of this report, the following general definition of ground treatment is suggested:

> Ground treatment is the controlled alteration of the state, nature or mass behaviour of ground materials in order to achieve an intended satisfactory response to existing or projected environmental and engineering actions.

1.3 PRINCIPAL METHODS FOR GROUND IMPROVEMENT

Ground treatment techniques have developed greatly over the past 30 or so years and the possibilities for new applications appear to be increasing. This part of ground engineering is one where practice precedes theory. Research follows development, not just in trying to explain why a technique works but, more importantly, to establish rational, rather than empirical, design methods and to see how the technique can be improved and its limitations identified. In the United Kingdom, some 75 per cent of the ground improvement contracts using the techniques of vibro-replacement and dynamic compaction are for man-made ground. These two techniques, including their application to loose or soft natural soils, are probably the commonest type of ground treatment used in the UK. For overseas work, the proportions of specialist ground treatments are reversed, ie 30 per cent are for man-made ground and 70 per cent for natural ground.

There are, of course, many other techniques than the two mainly used in UK. In this report, more than 30 are described separately, but for most there are variations that extend the versatility of the basic technique. Increasingly, different techniques are being used in conjunction (see Section 12.1) to considerable advantage in time and cost.

The techniques of ground improvement have been grouped into broad categories:

- improvement by vibration (Section 4)
- improvement by adding load (or increasing the effective stresses) (Section 5)
- improvement by structural reinforcement (Section 6)
- improvement by structural fill (Section 7)
- improvement by admixtures (Section 8)
- improvement by grouting (Section 9)
- improvement by thermal stabilisation (Section 10)
- improvement by vegetation (Section 11).

Although the above headings for the groups of methods reflect what is being done to the ground to improve it, they do not characterise the way the ground is to be improved, nor do they show the purpose of the improvement.

Many of the techniques can be used for different purposes and by enhancing one aspect of soil behaviour other aspects are also improved. The use of vertical drains accelerates the gain in strength of a soft foundation and allows settlement to take place more rapidly. The purpose of permeation grouting of a water-bearing sand to be tunnelled through would be to achieve sufficient strength for safe excavation; the complementary result of low permeability and minimal water inflow is a necessary though subsidiary benefit. This latter example highlights the need for thought about the purpose of the ground treatment. The problem is not the quantity of water that would enter the tunnel, but the instability of the ground, ie the sand would run with disastrous consequences. The treatment is not to achieve a low permeability but to create a uniform mass of sufficiently strong material able to resist the water pressure and prevent piping and erosion.

1.4 PURPOSES AND EFFECTS OF GROUND TREATMENT

Ground treatments should not just be thought of as temporary construction expedients, although many of the techniques are used to great advantage in this way. Increasingly, the improved ground is an integral part of the finished works, eg reinforced soil, stone columns by vibro-replacement. Even techniques intended as short-term improvement contribute to the permanent works. The use of vertical drains to speed up consolidation achieves a permanent improvement and the process is usually integrated with the construction of the permanent works. Three situations can be considered in relation to the purpose of the treatment:

- temporary, eg dewatering or ground freezing, where the improvement is only during the application
- short-term, eg some forms of grouting, or the use of basal reinforcement for an embankment on a soft foundation, where the treatment has a lasting effect, but its purpose is achieved during construction
- long-term, eg soil nailing, vibro-replacement, curtain grouting of a dam, where the treatment is integral to the permanent works.

The effects of ground treatment can be considered from two angles:

- the benefit or effect on the work of construction for which the treatment is sought, eg less or faster settlement
- the effect on the properties or behaviour of the ground, eg greater strength or stiffness.

The first of these is the driving force or justification for the treatment; the second is the identification of what aspect of the ground should be improved. Cost and practicability underlie both ways of defining the purpose of the treatment. Different techniques by different methods may achieve similar effects; usually they would have to be compared against a construction method not relying on ground improvement.

Table 1.1 *Benefits to construction work of different ground treatment techniques*

Technique	Benefit							
	Higher bearing capacity	Less or more even settlement	Faster settlement time	Ground-water control	Reduced liquefaction potential	Increased erosion resistance	Improved face/slope stability	Report section
Vibration								**4**
vibro-compaction	•••	•••			•••		•	4.1
vibro-replacement	•••	•••	•		•••		•	4.2
dynamic compaction	•••	•••			•••		•	4.3
vibratory probing	•••	•••			•••			4.4
compaction piles	•••	•••			•			4.5
blasting	•••	•••			•			4.6
Adding load								**5**
pre-compression	•••	•••						5.1
vertical drains	•••	•••	•••					5.2
inundation		•••						5.3
vacuum preloading	•••	•••						5.4
dewatering fine soils	•	•		•••	•	•	•••	5.5
pressure berms							•••	5.6
Structural reinforcement								**6**
reinforced soil	•	•				•	•••	6.1
soil nailing						•	•••	6.2
root and micro-piles	•••	•					•	6.3
slope dowels							•••	6.4
embankment piles	•	•••	•					6.5
Structural fill								**7**
remove-and-replace	•	•••	•					7.1
displacement	•••	•	•					7.2
reduced load		•••						7.3
Admixtures								**8**
lime/cement columns	•••	•	•	•••	•		•	8.1, 8.2
mix-in-place by single auger	•	•	•	•••	•		•••	8.3
lime stabilisation of slopes						•	•••	8.4
stabilisation of subgrades	•••	•••						8.5, 8.6
Grouting								**9**
permeation	•	•		•••		•	•••	9.2
hydrofracture	•	•••	•••	•			•••	9.3
jet grouting	•••	•••		•••			•••	9.4
compaction grouting	•••	•••	•					9.5
cavity filling	•	•••		•				9.6
Other methods								**10**
freezing				•••			•••	10.1
heating	•						•	10.2
vegetation							•	**11**

Key
Main benefit or purpose •••   Associated benefit or possible application •

In terms of benefit to the construction work, the main ground improvement objectives relate to three aspects:

- reduction and control of deformation
- reduction, control and exclusion of groundwater
- reduction of susceptibility to erosion.

In terms of effect on the ground, there are three types of change:

- change of state, ie the same ground but made stronger, stiffer, denser, more durable
- change of nature, ie the ground becomes a different material by inclusion of other materials
- change of response, ie through the incorporation of other materials, the ground becomes a composite material with enhanced load-carrying or deformation characteristics.

Table 1.1 lists the individual ground treatment techniques described in this book by group and in terms of their usual purpose or benefit for the construction works. The main purposes of ground treatment include:

- improving the bearing capacity of the ground
- reducing the potential for total and differential settlement (ie settlement management)
- reducing the time during which the settlement takes place
- reducing potential for liquefaction in saturated fine sands or hydraulic fills
- reducing the permeability of the ground
- removing or excluding water from the ground
- increasing the shear strength of the ground, and improving slope stability
- increasing the erosion resistance
- providing internal drainage systems in the ground (or vents for gas emissions).

It is suggested that as a start when considering ground treatment as an option, three questions should be posed:

- what are the required benefits to the main project?
- is the purpose a temporary, short-term or long-term improvement?
- what change in the ground is needed to achieve the benefit?

Ground treatment will normally be chosen as a compromise where shallow foundations cannot provide the required settlement performance and piled foundations are judged to be uneconomical; in other words, it is often adopted as a cheaper but less reliable solution than piling. The design and construction process for ground treatment should therefore incorporate sufficient checks and tests to establish that the required performance has been provided but without increasing costs to make piling the more preferable option.

1.5 REFERENCES

Charles, J A (1993)
Building on fill: geotechnical aspects
Report BR230, BRE, Garston, Watford

Holtz, R D, Jamiolkowski, M B, Lancellotta, R and Pedroni, R (1991)
Prefabricated vertical drains: design and performance
Book 11, CIRIA, London, and Butterworth-Heinemann, Oxford

Jewell, R A (1996)
Soil reinforcement by geotextiles
Special Publication 123, CIRIA, London, and Thomas Telford, London

Moseley, M P (ed) (1993)
Ground improvement
Blackie, Glasgow

Preene, M, Roberts, T O L, Powrie, W and Dyer, M R (2000)
Groundwater control: design and practice
Publication C515, CIRIA, London

Rawlings, G, Hellawell, E E and Kilkenny, W M (2000)
Grouting for ground engineering
Publication C514, CIRIA, London

Van Impe, W F (1989)
Soil improvement techniques and their evolution
Balkema, Rotterdam

2 Responsibility for design

Since the preparation of this report, the Construction (Design and Management) Regulations 1994 (CDM) have been instituted. The Regulations place additional responsibilities on all members of the design and construction team. Guidance on the application of CDM is given in CIRIA Report 166.

In Section 2.2, and specifically Table 2.2 below, there is suggestion for a division of responsibilities between the designer and specialist contractor. Other divisions are possible, and may be as valid, but what is most important is that risks and responsibilities are explicitly allocated, ideally to the party best placed to control them. To do this effectively, a party other than the specialist contractor should be involved to advise the promoter, procuring organisation, or specifier on achieving an equitable balance of risk.

For all projects where ground improvement is contemplated, there should be a clear understanding of the division responsibilities between those involved. This is crucial when neither promoter nor principal adviser has sufficient geotechnical expertise to design the improvement works. Equally, there should be no misunderstanding about what can realistically be achieved by the ground treatment process.

The roles of the promoter, the ground improvement designer and the specialist contractor in a ground improvement scheme are considered in the following sections.

2.1 PROMOTER

From a survey of more than 5000 construction projects, NEDO (1985) published *Thinking about Building*. This guide, for promoters of construction projects, is equally relevant to the promoter who is considering the risks of a site with either man-made or natural ground and whether or not ground improvement is needed, or can be beneficial. NEDO suggests that to achieve successful completion of the construction project a promoter should:

- delegate to an in-house executive sufficient power to act as promoter's representative for the project
- appoint a principal adviser – in effect the lead design professional for the project, who could also be the ground improvement designer
- carefully define the requirements for the project ie the brief
- make a realistic determination of project timing
- select an appropriate procurement path
- consider the choice of organisations to be employed other than the principal adviser
- have the site professionally appraised before being finally committed to it. This should always include a desk study.

These seven actions to be taken by the promoter are shown in Table 2.1, the flow path of the design process. Chapter 4.6 of NHBC Standards, on vibro-replacement, published by the National House-Building Council (1995), follows this approach.

Included in the NEDO guidance is advice to help promoters identify their priorities so that they can choose an appropriate procurement path for their project.

Two key questions concerning responsibility are:

- could the promoter manage separate consultancies and contractors, or is just one firm to be responsible after the briefing stage?

- would the professional responsibility of the designers and cost consultants be direct to the promoter?

It is these two questions which are discussed below in relation to the design of ground improvement. In this discussion, the term "designer" refers to the designer of the ground improvement scheme; as such the designer could be employed direct by the promoter, by a main works contractor, by a specialist contractor, or by a firm of consulting engineers. The designer could also be the principal adviser of the promoter. The designer's position in relation to the promoter depends, therefore, on the chosen procurement path. Before deciding which procurement path is to be adopted, the promoter should also consider the guidance given by the British Property Federation (1983) and the Institution of Structural Engineers (1984). The approach proposed by the British Property Foundation is intended to achieve speedier overall completion times. CIRIA Special Publication 15 (CIRIA, 1985) also gives guidance on the design-and-build approach to procurement.

Table 2.1 *The design and communication process for ground improvement*

Promoter	Ground treatment designer
• Considers site for project	
• Appoints promoter's representative	
• Appoints principal adviser	
• Determines project brief and timing	• Investigates site and ground
• Has proposals appraised professionally	• Establishes agreed design criteria
• Decides to proceed with project	• Evaluates engineering problems
	• Compares technical solutions with (and without) ground improvement
• Chooses procurement method	
	• Chooses treatment method after discussions with specialist contractors
• Agrees to proposal	
	• Prepares contract documents
• Places contract	
	• Supervises construction, testing and monitoring
• Receives technical and contractual reports	• Monitors performance of completed works
• Permits access and receives reports of performance	

To understand the promoter's point of view, designers will find useful advice in Kingsman (1985). He describes how commercial developers view problems with buildings including, as he named them, leaks, cracks, things that fall off, and mechanical services that do not work properly. Kingsman points out that commercial developers tend to be traditionalists on design matters rather than pioneers. Yet despite ground engineering being a continuous process of innovation, ground treatment is frequently used on industrial and commercial estate developments; for the promoters and designers of these types of project, MacCalman *et al* (1986) give useful guidance.

An alternative approach is for the promoter to take out a project-specific decennial insurance policy that provides cover against defects arising in the first 10 years of the life of a building. This is a form of "no fault" insurance for the professional team.

All parties to a ground improvement scheme would benefit from reading the 1988 NEDO report *Faster Building for Commerce*, which assessed more than 8000 projects. It discusses the effectiveness of the various procurement paths. The roles of the several members of the design team, of the client and of the contractors, are also appraised in detail. We are in an atmosphere of rapidly changing attitudes, methods, and lines of responsibility. NEDO (1988) advises that "the key to better performance is the coherent management of the design phase". There should be no ambiguity about leadership or the distribution of duties and responsibility under the contract and the law. How ambiguity in ground improvement schemes can be avoided is discussed in the following sections.

Table 2.2 *Responsibilities for design and construction of ground treatment works*

Designer	Specialist contractor
Obtains and interprets site and ground investigation data	Provides materials and equipment for improving the ground
Selects ground treatment method	Plans the working arrangements, sequence and operations within the specified design
Decides on need for trials	
Decides on testing strategy	Sets out and executes the ground treatment works
Assesses risks of scheme in such matters as noise, vibration and impact damage, and obtains details of adjacent structures	Tests and monitors the construction operations for internal quality control purposes
Assesses compatibility of the design of the proposed structure with the expected behaviour of the improved ground and the programme of the project, and assesses overall stability	Carries out tests and monitoring specified as a contractual requirement, which may include noise and vibration measurements
Prepares specification and contract documents	Completes the ground treatment works including additional or extra works instructed under the contract
Supervises the ground treatment works, including inspection by sampling and testing	
Advises on the need to vary the ground treatment	
Monitors the behaviour of the structure built on the treated ground	

2.2 DESIGNER

In addition to the relationship between the designer and the promoter through the design process illustrated in Table 2.1, there is consequent linkage between the responsibilities of designer and specialist contractor. Table 2.2 suggests a division of these responsibilities and duties between them. The relationships are put forward in this way to help avoid ambiguity about responsibility for design.

When the project brief is agreed between the promoter and the designer, both parties should be clear about the objectives of the ground treatment. A frequent purpose (using a building project as an example) is that of "settlement management" as and where poor ground is made stiffer. Treatment may be in specific areas only. Settlements are reduced but not to the extent that foundations would behave rigidly. For example, stone columns do not behave in the same manner or have the same stiffness as piles. This is illustrated on Figure 2.1, adapted from Meyerhof (1984). No treatment at all is one extreme, and piling the other: the results of ground treatment processes lie between them.

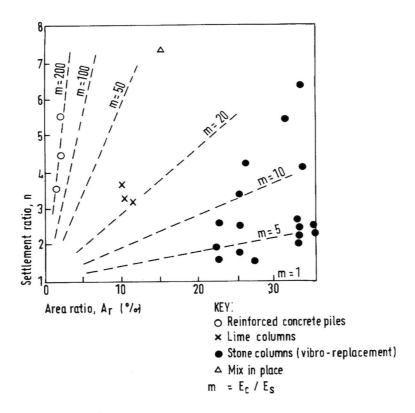

Figure 2.1 *Comparisons of settlement observations of foundations on piles and on improved ground (after Meyerhof, 1984)*

The designer has to set realistic objectives for the chosen scheme of treatment which the promoter understands and accepts, because, as Uff (1986) pointed out, increasingly courts are finding designers liable. The designer is advised to confirm the feasibility of the ground improvement targets in discussion with specialist contractors. The designer should also carry sufficient professional indemnity insurance against being held responsible for the failure of any ground treatment works. When the concept of the "design" has been finalised it is essential that it is communicated to both the main and specialist contractors. What is critical, or risky, should be clearly explained.

Table 2.1 shows the designer as responsible for the design of the ground treatment: these responsibilities can be delegated to the specialist contractor, but only with the written agreement of the promoter (Uff, 1986; Cornes, 1989). They must then be clearly reflected and defined in the contract documents.

2.3   SPECIALIST CONTRACTOR

The suggested responsibilities of the specialist contractor are shown in Table 2.1. If the promoter delegates the design responsibility to the specialist contractor to include site investigation and establishing the treatment to be used for the project, the specialist contractor should carry suitable professional indemnity or product insurance. The promoter and the specialist contractor with design responsibility also need to be clear about the extent of their responsibilities and duties in the process shown in Figure 2.1.

This applies equally when a main contractor employs the specialist contractor on behalf of the promoter, and when there is a combination of techniques. An example of this is the use of precompression filling where the earthworks are the main contractor's responsibility, but the vertical drains are the responsibility of the specialist contractor.

2.4   REFERENCES

British Property Federation (1983)
Manual of the British Property Federation System for building design and construction,
BPF, London

CIRIA (1985)
A client's guide to design and build
Special Publication 15, CIRIA, London

CIRIA (1997)
CDM Regulations – work sector guidance for designers
Report 166, CIRIA, London

Cornes, D L (1989)
Design liability in the construction industry
BSP Professional Books, Oxford, 3rd edn

Institution of Structural Engineers (1984)
Guide to the performance of building structures
Instn Struct Engrs, London

Kingsman, C (1985)
"Building problems and design considerations: the commercial developer's viewpoint"
In: *Design life of buildings*
Thomas Telford, London, pp 113–120

McCalman, C D, Bennett, R C, Griffiths, R, Hall, J W, Mann, D H, Pavitt, J H, Skyrme,
V H and Stevenson, M W (1986)
Industrial and commercial estates: planning and site development
Thomas Telford, London

Meyerhof, G G (1984)
Closing address in: *Piling and ground treatment*
Thomas Telford, London, pp 293–297

National Economic Development Office (NEDO) (1985)
*Thinking about building: a successful business customer's guide to using the
construction industry*
National Economic Development Office, London

NEDO (1988)
Faster building for commerce
National Economic Development Office, London

National House-Building Council (NHBC) (1995)
"Vibratory ground improvement techniques"
Chapter 4.6 in *NHBC Standards*, April, National House-Building Council, London

Uff, J F (1986)
The impact of Common Law on the Engineer's Duty
Proc Instn Civ Engrs, Part 1, Vol 80, June, pp 807–809

3 The design process

The designer of a ground improvement scheme has to consider the use of possibly several techniques offering viable solutions for the project. In many circumstances it will be necessary to compare total project costs to select the most appropriate ground improvement scheme, which might not necessarily be the cheapest on its prime cost alone. The compatibility of the structure and the degree of improvement that is given to the ground, the ancillary works – which include control and performance testing – and the overall project programme are among the factors that have to be considered in devising such a cost comparison.

The following sections deal with the process of the design of ground treatment works and discuss particular aspects that the designer should take into account.

3.1 SITE INVESTIGATION

The documents produced by the Site Investigation Steering Group (Thomas Telford, 1993) provide guidance for clients and other members of the professional team on the benefits of, and how to procure, proper site investigation.

Faster building for commerce (NEDO, 1988) stated that unexpected ground conditions delayed one in two projects. The immediate inference is that many site investigations are inadequate, despite efforts in recent years to improve standards and to increase client awareness of the value of thorough ground investigation. The reality is that the ground remains one of the greatest sources of risk for all construction works.

Successful treatment is where the ground is improved so as to reduce some of these risks to acceptable levels, but this does not mean investigating the ground is any the less important. Rather, the investigation should be designed to provide information relevant to the contemplated techniques of ground improvement. Often this requires detailed knowledge of the ground structure, eg of the macrofabric, when vertical drains or permeation grouting might be needed, and extensive field testing as a baseline for evaluating the degree of improvement achieved by the treatment.

All sites should be investigated in accordance with BS 5930: *Code of Practice for Site Investigation*. Site investigation involves much more than ground investigation. It is a continuing process, although it is convenient to identify five stages:

- desk study
- site reconnaissance
- detailed examination for design and for safety during construction, which includes ground investigation, topographic and hydrographic surveys, and of adjacent properties, and special studies, eg effects of vibration, contamination, toxic waste and gases. This work can be carried out in several stages
- investigation during construction
- monitoring the performance of the structure.

The designer and a geotechnical specialist should plan carefully the extent and sequence of the site investigation (Uff and Clayton, 1988; BRE, 1987b). As BS 5930:1999 points out: "The imposition (for reasons of cost and time) of limitations on the amount of ground investigation to be undertaken may result in insufficient information being obtained to enable the works to be designed, tendered for and constructed adequately, economically and on time. Additional investigations carried out at a later stage may prove more costly and result in delays."

Figure 3.1, from CIRIA Special Publication 25 *Site investigation manual* (Weltman and Head, 1983), illustrates a suitable sequence for a site investigation. Henkel (1984) confirms that this sequence is appropriate to ground improvement projects. Published as a draft for development, DD 175: 1988, *Draft Code of Practice for the identification of potentially contaminated sites and their investigation* (BSI, 1988, in revision) is an essential reference to be consulted for any site that includes derelict or previously used land. For further guidance on site investigation, readers should refer to the series of publications (referenced at the end of Chapter 3) of the broadly based Site Investigation Steering Group and to Clayton *et al* (1995).

3.1.1 Desk study

Desk study is the term used to describe the collection and examination of information about the site and its surroundings. As well as geological information and reports of previous investigations, relevant information can be gained from published material, old maps, and archive records about the history and past use of the site, about groundwater conditions and many other matters. The desk study is a continuing search for information. Guidance on sources of information is given by Dumbleton and West (1976), by Dumbleton (1980) regarding historical searches for sites, by BS 5930:1999, by Weltman and Head (1983), and by examples in BRE (1987a and b). Information sources for areas of industrial development can be found in BS 5930 and in Healy and Head (1984). In addition, Gutt *et al* (1974) presented a survey of the locations, quantities, and methods of disposal or use, of all the major waste materials in England and Wales. Corbett (1982) gives sources of information about the legacy of wars, such as First World War factories for filling shells with phosgene and mustard gas.

Aerial photographs can be a powerful way of assessing a site, particularly when they show changes in the site from earlier photographs. Practical help, related to roads, is given by Dumbleton and West (1970), and by Dumbleton (1983) for engineering sites in general. This second reference notes 20 major sources and collections of aerial photos.

Figure 3.2 illustrates the value of aerial photography. The plan of two buildings has been superimposed on the aerial photograph of the site prior to redevelopment. About 0.8 m settlement occurred in the pond area, which had been filled shortly before construction. The remainder of the building area settled only slightly because the area had been filled more than 30 years earlier. The aerial photographs thus provided an explanation for the great variation in observed behaviour between different parts of the buildings.

Guidance about methane in relation to derelict land is given by Leach and Goodger (1991), and in the series of reports of CIRIA's programme "Methane and associated hazards to construction", ie Hartless (1992), Hooker and Bannon (1993), Crowhurst and Manchester (1993), and Card (1993). These cover the occurrence and measurement of methane, investigation procedures and protection measures.

Infra-red photography, using model aircraft to mount cameras, can also be used to investigate migration of methane beyond site boundaries. The method of investigation is described by Weltman (1983) and Whitelaw (1986).

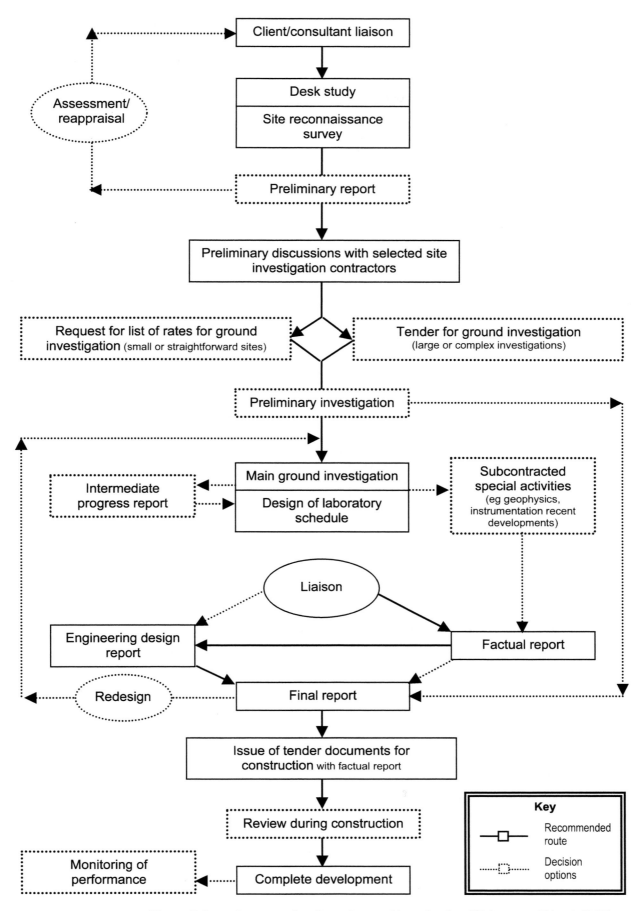

Figure 3.1 *Sequence for site and ground investigation (Weltman and Head, 1983)*

The aerial photograph was taken on 31 May 1960. Superimposed are the outlines of two buildings – part of a major redevelopment in this area. These buildings experienced severe differential settlement, up to 0.8 m where the pond had been. The site of the pond can be seen as the darker triangle towards the right-hand side of the building area. It had been backfilled not long before construction of the buildings.

Figure 3.2   *Example of the usefulness of aerial photographic interpretation*

3.1.2   Site reconnaissance

Guidance on site reconnaissance is given in BS 5930:1999 and by Weltman and Head (1983). As well as taking a thorough look at what exists on the project site, the surroundings – particularly the nearby structures and their usage – should be inspected. Not only will these observations influence the direction of the desk study and ground investigation, but also they are particularly relevant to the choice of a ground treatment method, and the need to consider environmental constraints.

3.1.3 Ground investigation

Whenever possible, the ground investigation should be conducted in stages. The first stage should permit the geology and groundwater of the site to be understood properly. Subsequent stages may include investigation of the likely design options, such as shallow foundations, piling or different ground treatment techniques. *In-situ* testing as part of the ground investigation should be similar to the testing that is to be used later in assessing the degree of improvement achieved. Table 3.1 summarises the geotechnical information typically required for the design of ground treatment schemes. Eakin and Crowther (1985) point out that it may be possible to record the position of buried obstructions during the demolition contract or, as Mitchell (1987) found, during the cut-and-fill operations of site preparation prior to the ground treatment contract.

3.2 ASSESSING POSSIBLE RISKS IN GROUND IMPROVEMENT PROJECTS

3.2.1 Hazards

It has been noted that some 75 per cent of treatment projects by vibro-replacement and dynamic compaction in the UK involve man-made ground, the diverse nature of which can pose hazards to development. Seago and Treharne (1985) list many of the following potential hazards of made ground:

- excessive settlement, which will not be uniform
- gas generation, usually of methane and hydrogen sulphide, and carbon dioxide
- chemical attack on buried things
- chemical attack on living things
- combustion, eg of coal wastes and shales
- contamination of environment, eg in leachates or in dust.

Even where the ground treatment is for a development on soft or loose natural ground, the first three potential hazards have to be considered.

In each treatment, the risks associated with these potential hazards need to be assessed, not only for the stage when the whole project is complete, but also for the separate stages of construction. Further guidance on the subject of development on filled and contaminated land is given in *Proceedings of the Glasgow Conference on engineering on marginal land* (Thomas Telford, 1987) and in Leach and Goodger (1991).

3.2.2 The promoter

Table 2.2 indicates that the promoter should carry out a site investigation before purchasing a site. This allows the one person who has the entire project picture to carry out an informed assessment of risk.

Where, for example, a development has to include some more heavily loaded structures, these can be located on the better ground on the site, avoiding the need to improve the ground. The promoter can then consider whether to develop the site in a way that permits a later change of use or to go ahead with a specific form of development. The latter is usually more economical.

Table 3.1   *Geotechnical information typically needed for different treatments*

Intended treatment	for the site: S			for treatment target zone: T		for comparative observations: O			
	Ground profile/ structure	Groundwater conditions	Obstructions	Fabric description	Classification tests	Shear strength	Compressibility	Permeability	Relative compaction
Vibration									
vibro-compaction	ST	STO	S	ST	ST				TO
vibro-replacement	ST	STO	S	ST	ST	ST			TO
dynamic compaction	ST	STO	S	ST	ST				TO
vibratory probing	ST	STO	S	ST	ST				TO
compaction piles	ST	ST	S		ST				TO
blasting	ST	STO	S		ST				TO
Adding load									
pre-compression	ST	STO		ST	ST	STO	STO	STO	
vertical drains	ST	STO	S	ST	ST	STO	STO	STO	S
inundation	ST	STO		ST	ST	ST	STO	S	
vacuum pre-loading	ST	STO		ST	ST	STO	STO	STO	
dewatering fine soils	ST	STO	S	STO	ST	ST	S	STO	
pressure berms	ST	STO		ST	ST	STO	STO		
Structural reinforcement									
reinforced soil	ST	ST			S	S	S		
soil nailing	ST	ST	S	ST	ST	ST			
root and micro-piles	ST	ST	S		ST	ST			
slope dowels	ST	ST	S		ST	ST			
embankment piles	ST	ST	S		ST	ST			
Structural fill									
remove and replace	ST	ST	S		ST	ST	S		
displacement	ST	ST	S		ST	ST	ST		
reduced load	ST	ST		ST	ST	ST	ST		
Admixtures									
lime/cement columns	ST	ST	S	S	ST	STO	ST	STO	
mix-in-place by single auger	ST	ST	S	S	ST	STO	ST		
lime stabilisation of slopes	ST	ST	S	ST	ST	STO			
stabilisation of subgrades	ST	ST	S		ST	STO			
Grouting									
permeation	ST	STO	S	ST	ST	S		STO	
hydrofracture	ST	STO	S	ST	ST	S		STO	
jet grouting	ST	ST	S		ST	STO		STO	
compaction grouting	ST	STO	S	ST	ST	S	S	STO	TO
cavity filling	ST	STO	S					S	TO
Other methods									
freezing	ST	STO	S	S	ST	S		STO	
heating	ST	ST	S	S	ST	S		STO	
vegetation	ST	ST			ST			TO	

Table 3.2 summarises the key questions the promoter should address. The answers bear heavily on the brief for the scheme of ground improvement and the risks to be considered. Timing (A in Table 3.2) is an important factor. A short programme may appear to rule out pre-loading, which for a large enough site, and with more time, could be an obvious option. Even when time is short this may prove an economical solution, because, as Charles *et al* (1986) pointed out, most of the settlement of unsaturated loose fills occurs as the surcharge is placed upon it. Tomlinson and Wilson (1973) observed the same for colliery waste.

Table 3.2   *Key questions for the promoter (adapted from* Thinking about building, *NEDO, 1985)*

A. Timing	How important is early completion to the success of the project?
B. Controllable variation	Will there be a need to alter the project once it has begun on site, eg by changing machinery layouts?
C. Complexity	Does the structure itself (as opposed to its contents) have to be technically advanced or highly serviced?
D. Quality level	What level of quality in design and workmanship is required? Does the building have to be rigid?
E. Price certainty	Is a firm price needed before commitment to proceed with the project?
F. Competition	Can the construction team be chosen by value rather than price competition?
G. Responsibility	1. Would the promoter prefer to manage separate consultancies and contractors, or to give single-point responsibility to one organisation?
	2. Would the professional responsibility of the designers and cost consultants be direct to the promoter?
H. Risk avoidance	Would the promoter pay someone to take the risk of cost and time stoppage?

Questions B, C and D of Table 3.2 relate to the user's requirements from the development. In many cases, an important matter to establish is the specification for the ground slab of the building(s) on the development. Deacon (1987) gives detailed guidance on the design of concrete ground floors. A realistic estimate of loading has to be made, perhaps planning for greater loadings should the structure be resold. Further guidance on the determination of appropriate loadings for warehouse floors is given by Armitage and Judge (1987) and Neal and Judge (1987). Current guidance on floor loads for use in design is provided by the code of practice BS 6399: Part 1:1996 *Loading for building – code of practice for dead and imposed loads.*

The required quality level of construction and finishing for the new buildings (D in Table 3.2) affects the choice and design of the ground treatment scheme. Where a high quality of finish using brittle materials is required, or where a very sensitive piece of equipment or plant is to be installed, piling, rather than ground treatment, would probably be necessary. For structures, ground treatment techniques typically improve the stiffness of poor ground by between two and three times (see Figure 2.2) – up to three to four times or even higher can be achieved in some granular soils. Ground and settlement variability is also reduced. Successful ground improvement achieves a match between the ground conditions and the requirements of structural performance. Table 3.3 from Mitchell (1986) is a useful guide, because it relates structure type to ground where treatment is likely to be advantageous. In the same context, it can be said that the value of treatment rises as load intensity decreases and the size of loaded area increases.

In a paper more related to the likely views of a promoter, Cragg and Walker (1987) suggest that it is usually market conditions that determine project viability, rather than ground conditions. They describe, with five examples, how a promoter may choose to accept, or avoid, risks in developing sites with poor ground. Skilton (1987) gives a further example, where a promoter agreed to the use of a combination of ground treatment with jackable foundations, instead of piles. It is therefore essential that the promoter be correctly advised by the designer as to the level of performance possible with each improvement process.

Table 3.3　*Applicability of foundation soil improvement for different structures and soil types (from Mitchell, 1986)*

Category of structure	Structure	Permissible settlement	Load intensity/ usual pressure required, kPa	Probability of advantageous use of soil improvement techniques in:		
				Loose cohesionless soils	Soft alluvial deposits	Old, uncontrolled inorganic fills
Office/ apartment	High-rise: more than six storeys	Small < 25–50 mm	High/300+	High	Unlikely	Low
Frame or load-bearing construction	Medium-rise: 3–6 storeys	Small < 25–50 mm	Moderate/200	High	Low	Good
	Low-rise: 1–3 storeys	Small < 25–50 mm	Low/100–200	High	Good	High
Industrial	Large span with heavy machines, cranes; process plants; power plants	Small (< 25–50 mm) Differential settlement is critical	Variable/ high local concentrations to > 400	High	Unlikely	Low
	Framed warehouses and factories	Moderate	Low/100–200	High	Good	High
	Covered storage, storage rack systems, production areas	Low to moderate	Low/< 200	High	Good	High
Others	Water and wastewater treatment plants	Moderate. Differential settlement is important	Low/<150	High*	High	High
	Storage tanks	Moderate to high, but differential may be critical	High/up to 300	High*	High	High
	Open storage areas	High	High/up to 300	High*	High	High
	Embankments and abutments	Moderate to high	High/up to 200	High*	High	High

* Improvement likely to be required for liquefaction control in seismic areas.

3.2.3 The designer

Once sure that the promoter's requirements are fully understood, the designer can then draw attention to the risks. The designer should expect the promoter to test the need for, and the strength of, the design recommendations. Any risks inherent in the designer's proposals have to be accepted by the promoter.

Typical engineering questions for the designer to address are summarised in Table 3.4, which is based on Charles (1984) and Eakin and Crowther (1985).

Table 3.4   *Engineering risks in ground improvement*

- settlement under self-weight (creep settlement)
- settlement under increased load
- settlement because of inundation
- consolidation of fine-grained materials
- toxicity of wastes
- variability of wastes
- old retaining walls and foundations
- mining subsidence
- expansive steel slags
- underground obstructions

- cavities or voided ground
- fire from spontaneous combustion
- explosion from methane gas emissions
- stability of sloping ground
- changes to surface drainage regimes
- changes to groundwater conditions (eg drainage, changes of level)
- influences of other construction
- durability
- health and safety considerations

In many ground improvement projects the primary purpose is settlement management of loose fills and made ground. Much of the fill remains as compressible (or nearly so) as it was before ground treatment. It is necessary to assess both the potential for settlement and the causes of settlement. For fills, there are several aspects to be considered:

- settlement under self-weight (creep settlement)
- settlement under increased load
- settlement because of inundation (collapse settlement), which may be because of a rise in groundwater level or the entry of rainwater
- settlement by consolidation, eg of fills placed under water.

These are discussed in Appendix A.

3.3   CHOICE OF TREATMENT TECHNIQUE

After assessing the hazards and risks (Section 3.2) and agreeing with the promoter that ground improvement is consistent with the building function, the designer can choose an appropriate technique. Figure 3.3 is an example of the options for foundation solutions. In selecting a method of ground treatment, Mitchell (1982) suggests that the following factors should be taken into account: those in italics are the most important.

- the *purpose* of the ground treatment, to establish the level of improvement required in terms of ground properties such as strength, stiffness, compressibility and permeability
- the area, depth and total volume of soil to be treated
- soil type and its initial properties
- materials availability, eg sand, gravel, water and mixtures
- availability of equipment and *skilled personnel*
- environmental factors, such as waste disposal, erosion, water pollution, effects on adjacent structures and facilities
- local experience and preferences
- time available
- cost.

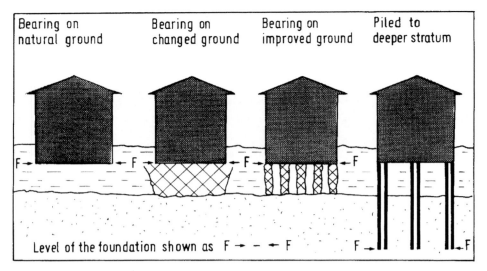

Figure 3.3   *Foundation options*

The principal treatment methods are listed in Table 1.1 and described in Chapters 4 to 11 in this book. (These descriptions include variations of the principal techniques.) Further discussion on choice can be found in Chapter 12, and notes of some comparative case histories in Appendix B.

In many cases, the end result of the treatment process is an enhancement of several properties. Stone columns, while acting to reduce differential and total settlements, may also provide drainage channels accelerating consolidation, or permit the dissipation of pore pressures generated by seismic loadings. They may also act as vents for gas emission. Vertical drains, in accelerating consolidation, increase the shear strength of the ground so that the bearing capacity is improved. Grouting affects a soil's permeability and its shear strength.

In some circumstances the primary objective of the ground improvement technique may mask less desirable effects. An example might be the inadvertent discharge of surface water into a stone column, thereby channelling the water to the poor ground at depth around the column. Grouting, successful in rendering the ground less permeable, might strengthen it so much that excavation through it becomes more difficult. In Singapore, jet grouting stabilisation in organic clays generated ammonia and greatly increased the ground temperature, hindering the tunnelling (Bell, 1988).

As well as choices between methods to improve the ground, there may be options that do not involve treating it. Designing the structure to accommodate large movements (Heathcote, 1965), and providing "floating foundations" (Golder 1964 and 1975) may be appropriate solutions for some projects.

3.4   DISCUSSION WITH SPECIALIST CONTRACTOR

Where more than one treatment method appears to be appropriate for the project, the designer should discuss the options with specialist contractors. The main aim is to check the compatibility of structural behaviour with what may realistically be expected from the method. These discussions should take place as early as possible in the feasibility stage of design to benefit from the experience of the specialists. In the UK, there are specialist contractors with more than 30 years of experience with vibro-compaction and vibro-replacement, and more than 20 years' experience of dynamic compaction. One specialist contractor has been carrying out vertical drain projects for over 30 years.

Designers should ask the specialist contractors for information from case histories of comparable schemes, including the results of full-scale or zone loading tests or performance observations. The data may then be compared with the functional requirements of the building to assess the appropriateness of any treatment process.

Observations of the behaviour of real structures constructed on improved ground would be of the greatest help to designers, but unfortunately they are too rarely made. Ideally, the ground improvement designer should undertake the measurements. When such information is available it is also of value to the promoter in assessing risks (Dale, 1987) or, later, if the site or structure is to be sold.

That so many structures are founded upon treated ground and so few case histories have been published suggests that the applications were generally successful. (About one-third of those that have been published, being mainly for oil storage tanks, are where proof-load monitoring is a usual contract requirement, ie for the structure, irrespective of whether or not the ground has been treated.) Although cases of inadequacy tend not to be published for legal and commercial reasons, the lessons that are learnt from them influence practice, though not in an overt way that would establish greater acceptance of the technique or confidence in its applicability limits. The benefits of monitoring are not only for the specific structure, but also for the greater understanding of the capabilities of the method.

3.5 TESTING PROGRAMME

As the basic material to be improved is the ground, which can be extremely variable, the designer has to consider how to demonstrate that the design assumptions will be fulfilled. An essential part of ground improvement design is therefore to plan a testing programme to provide:

- information about the initial condition of the ground (if not obtained in the investigation results used for the selection and design of the treatment scheme)

- construction controls during the course of the treatment works, developed after initial site trials where necessary

- comparative indicators of the improvement achieved

- observations of the environmental or other effects of the treatment

- confirmation that engineering design values have been achieved

- monitoring the structure built upon the treated ground.

Table 3.5 lists *in-situ* tests and observations that have been used for the design and control of ground treatment works. Tests before and after treatment are obviously essential. They should show that not only has the ground been changed but also, and perhaps more importantly, that its variability has been reduced. From careful observations during the initial treatments, inspections of trial excavations, and testing on trial areas – at large or full scale, where appropriate (eg see Greenwood, 1991) – the working methods and testing programme can be developed. Compliance criteria specified for the chosen test method should be related to the performance required (eg settlement tolerances) for the proposed structure.

Table 3.5 *Types of tests and observations used to control and assess ground treatment*

Tests and observations	Initial state	Construction controls	Comparator of change	Effects on environment	Design values
In-situ testing					
SPT	●		●		
CPT	●		●		
DPT	●				
Vane tests	●	●	●		●
Pressuremeter tests	●		●		
Dilatometer	●		●		
Permeability tests	●		●	●	●
Plate loading tests	●		●		●
Full-scale load tests	●		●		●
Site observations					
Surface levelling	●	●	●	●	●
Treatment exposure (trial pits)		●	●		
Block or tube sampling of treatment			●		●
Method controls					
Locations of treatment points		●			●
Quality of treatment materials		●	●		●
Quantity of treatment materials		●	●		●
Procedures		●		●	
Limiting parameters		●	●	●	●
Control instrumentation		●	●	●	
Instrumentation					
Pore pressures	●	●	●	●	●
Structure movements	●	●	●	●	●
Noise and vibration	●	●	●	●	●
Settlement at depth	●	●	●	●	●
Lateral ground movements	●	●	●	●	●

SPT – Standard Penetration Test
CPT – Cone penetration test
DPT – Dynamic penetrometer test

Full-scale loading tests on trial areas are the best way to assess the effectiveness of the treatment for isolated loaded areas. This applies to any treatment process where the ground is being improved to support load and where the load/settlement relation is critical to structural design. Figure 3.4 gives details of a test arrangement for loading bases up to about 3 m square. This method was used for the zone tests reported by Bell *et al* (1987), by Mitchell (1987) and by Slocombe (1989). Test pads of this size were also used by Somerville (1987), and Slocombe and Moseley (1987) to test the behaviour of dynamic compaction.

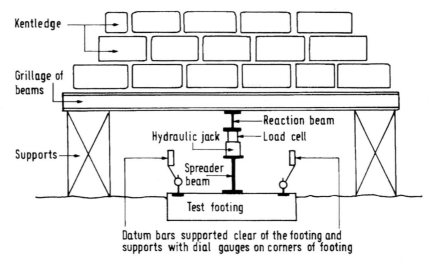

Figure 3.4   *Arrangement of a zone test*

A skip test may be used to examine the behaviour of an area of a ground slab or a lightly loaded strip footing on treated ground. Charles and Driscoll (1981) describe the test, and how it should be monitored for at least four weeks, to examine creep behaviour. Figure 3.5 shows a test arrangement and its results. BS 1377:1990:Part 9 includes the skip test as a shallow maintained-load test. McEntee (1987) describes how the skip, or skips, should be positioned on a concrete block. Levelling stations should be on the block and not the skip. A precise level must be used for determining settlements.

Vibro-replacement (stone columns) are often used to reduce liquefaction potential under seismic conditions as well as acting in the foundations of embankments and reinforced earth walls. Engelhardt and Golding (1976) measured the lateral resistance of stone columns – perhaps the only reported instance of this type of testing.

Note that CIRIA Research Project 604 about the engineering properties of treated ground offers guidance on the measurement of the properties and performance of different treatments. Where possible it presents case history information.

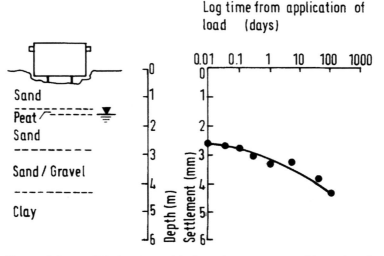

Figure 3.5   *Skip load test: (a) view of arrangement; (b) results of test on untreated ground*

3.6 CONTRACT DOCUMENTS

Model procedures and specifications for ground treatment have been published by the Institution of Civil Engineers (1987). That document describes the various methods of inviting ground improvement tenders and provides specification details for materials and workmanship for:

- vibro-compaction
- vibro-replacement/displacement (stone columns)
- dynamic compaction
- deep drains.

Reference should be made to the ICE Specifications in relation to scheduling testing for the treatment, recommended forms for bills of quantities, and for its discussion of design responsibility.

The document does not provide guidance on performance criteria. Determining realistic and appropriate criteria for the performance of the treatment – in effect trying to set an "end-product" specification – is fraught with difficulty. Certainly it should not be attempted solely on the basis of absolute values of index test results, such as the SPT, or in terms of degree of compaction, such as relative density (which is an attribute that cannot be measured in practice). This is not to say that these tests (before and after) cannot be used as a means of controlling the work. Swain and Holt (1987) describe some of the problems of using such a specification for dynamic compaction and of measuring completed work.

In-situ tests for ground treatment should be used either as a way of comparing relative improvement (eg penetration testing before and after treatment) or, for large-scale tests, as an indicator of the performance of the works for which the ground was treated (eg load/settlement behaviour of a structure).

It is essential, then, that the specialist contractor knows the performance criteria for the intended structure. Details of the structure itself should be provided, so that specialist contractors can make their own assessment of risk.

Qualifications of specialist contractors' tenders are usually stimulated by technical issues. These should be examined, discussed, appraised and, if possible, eliminated by agreement before the contract is made.

The documents should also describe a framework for changes in construction in the event of unexpected conditions being encountered on site. There should be suitable provision in the bill of quantities for these variations. Perhaps the most appropriate method is a specific bill item for supervision from both the designer and the specialist contractor. This would then satisfy the substantial "judgement on site" element associated with ground treatment processes. NHBC (1995) requires full-time supervision by the designer of vibro-replacement. This approach is to be pursued if at all possible.

3.7 SUPERVISING THE CONTRACT

Ground treatment is often a rapid process. On-the-spot control by the designers is essential to ensure a flexible and early response to unexpected or unforeseeable conditions. Uncertainties of the ground pose a risk for all construction works. The risks are first to the ground treatment itself – a change of techniques or additional work could be proportionately very expensive – and second to the construction project as a whole –

if the treatment proves less successful than intended or if there is delay. D'Appollonia (1984) discusses the questions of the unknowns in ground treatment.

An example might be stone columns having shallower depths than expected. Trial pits would then be necessary to establish the reason, eg the fill not being as deep as had been thought. Extra stone columns in the deeper-filled areas might then be needed to achieve comparable stiffness of the deep and shallow foundations. With equipment producing 30 to 100 stone columns per day, for example, such events require the designer to have suitable supervisory staff who can immediately adjust the form of the treatment, in conjunction with the specialist contractor, before the rig completes its work. The improvement process itself is thus a continuation of the site investigation, and the treatment is adjusted in accordance with the findings during the works.

Observation of the construction process is important so that a supervisor can assess whether, for example, a bottom-feed vibrator is feeding stone properly. *In-situ* penetration tests, vibration measurements and load tests should also be supervised. Such data can then be plotted immediately and compared with results of tests before treatment.

For most ground treatment processes, an observational approach (see Nicholson *et al*, 1999) is essential so that a judgement can be made as to what more is needed or as to how the process should be modified to increase efficiency. The observational method relies upon thorough records not only of what was done but also of difficulties or errors. Increasingly, instrumentation on rigs and batching systems, together with servo-systems controlled by computer, are improving the quality and amount of information about the completed work. Nevertheless, when the control of operations becomes more automated there is greater need for direct checks that the systems are working properly. Whether the records are in electric form or as handwritten logs, they should be available for checking and use during, and certainly immediately after, each treatment operation. Table 3.6 lists the items to be recorded for vibro-replacement and dynamic compaction.

The other element of the observational approach is choosing the indicators of satisfactory improvement, eg when grout take suggests that the ground has tightened up. On these matters, the experience of the specialist contractor is essential to achieve economy of treatment compatible with meeting acceptance criteria.

Health and safety aspects of the work must also be considered. Skipp and Hall (1982) in CIRIA Report 95 gave guidance on ground treatment materials, but this should be considered in relation to requirement of COSHH (Control of Substances Hazardous to Health) Regulations.

Table 3.6   *Routine monitoring of vibro-replacement and dynamic compaction (after Greenwood and Kirsch, 1984; Crossley and Thompson, 1987)*

Vibro-replacement	Dynamic compaction
• Location, date and time	• Location, date and time
• Depth	• Surface settlement and heave after each pass
• Consumption of stone per metre	• Imprint volumes and depths
• Speed of penetration and withdrawal of vibrator	• Numbers of blows
• Current or hydraulic oil pressure with depth	• Height of drop of pounder
• Plate tests for workmanship	
• Records of power consumption	

MONITORING PERFORMANCE

The only way for the designer to know that the end product has functioned as intended is to monitor the actual behaviour of the finished works. Often this is a building structure, for which settlement monitoring on the structure itself would be the best indicator of performance. In other circumstances, measuring the settlements of fills, the pore pressures in embankment foundations or those on either side of groundwater barriers, or ground movements at depth would be appropriate. As noted in Section 3.4, there is a lack of published, well-understood case histories.

Cheney (1973) describes how settlement monitoring can be carried out most effectively to accuracies better than 0.5 mm, using a surveyor's precise level. The Building Research Establishment (BRE) developed a suitable levelling station, shown in Figure 3.6. Details of how to install the socket are shown in the diagram. The critical point is that the loose thread draws the machined plane face of the levelling plug against that of the socket. It is thus a repeatable position, essential for accurate levelling.

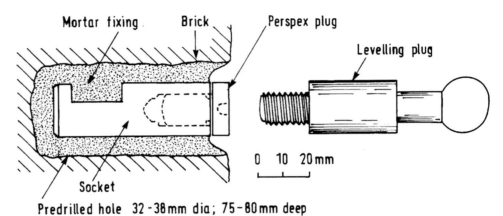

Figure 3.6   *BRE levelling station*

Provision of a suitable datum point is also essential for any levelling survey. Cheney describes details of a BRE 6 m-deep datum and its installation; an alternative datum is described by Burland and Moore (1973).

Movement at depth can also be monitored using the magnet extensometer of the kind described by Burland *et al* (1972). Charles *et al* (1986) reports several case histories where preloaded uncompacted fills were monitored with magnet extensometers. This system was used to measure the behaviour of fill improved by vibro-replacement (Skilton, 1987).

Observation wells to monitor groundwater may be needed, particularly for sites dewatered prior to dynamic compaction, or by earlier industrial use (see Appendix A, Section A.3 for the effects of inundation or rising groundwater). Permeability testing may be useful to monitor changes. Gas emissions may also require monitoring, using similar standpipe type installations. Details of a typical well are given in Crowhurst and Manchester (1993), who discuss methane monitoring in detail.

3.9 REFERENCES

Armitage, J S and Judge, C J (1987)
Floor loadings in warehouse: a review
BRE Report BR9, Building Research Establishment, Garston

Bell, A L (1988)
"Report of discussion on Coomber, D B on Jet Grouting"
Proc Instn Civ Engrs, Part 1, Vol 80, pp 1661–1664

Bell, A L, Kirkland, D A and Sinclair, A (1987)
"Vibro-replacement ground improvement at General Terminus Quay, Glasgow"
In: *Building on derelict and marginal land*, Proc Conf, Glasgow, 1986
Thomas Telford, London, pp 697–712

British Standards Institution (BSI)
BS 1377:1990 *Methods of testing soils for civil engineering purposes*
BS 5930:1981 *Code of Practice for site investigations*
BS 5930:1999 *Code of Practice for site investigations*
DD 175:1988 *Code of Practice for the identification of potentially contaminated land and its investigation*
BS 6399: Part 1:1996 *Loading for building – code of practice for dead and imposed loads*
BSI, London

Building Research Establishment (BRE) (1987a)
Site investigation for low-rise building: desk studies
Digest 318, BRE, Garston

BRE (1987b)
Site investigation for low-rise building: procurement
Digest 322, BRE, Garston

Burland, J B, Moore, J F A and Smith, P D K (1972)
A simple and precise borehole extensometer
Current Paper 11/72, Building Research Establishment, Watford

Burland, J B and Moore, J F A (1973)
"The measurement of ground displacement around deep excavations"
Proc Symp Field Instrumentation, Cambridge, British Geotech Soc
Butterworths, London, pp 70–84

Card, G B (1993)
Protecting development from methane
Report 149, CIRIA, London

Charles, J A (1984)
"Settlement of fill"
In: P B Attewell and R K Taylor (eds), *Ground Movements and their Effects on Structures*
Surrey University Press, Guildford, pp 26–45

Charles, J A and Driscoll, R M C (1981)
"A simple in-situ load test for shallow fill"
Ground Engineering, Vol 14, No 1, January, pp 31–36

Charles, J A, Burford, D and Watts, K S (1986)
"Improving the load-carrying characteristics of uncompacted fills"
Municipal Engineer, Vol 3, No 1, pp 1–19

Cheney, J E (1973)
"Techniques and equipment using a surveyor's level for accurate measurement of building movement"
Proc Symp Field Instrumentation, British Geotech Soc
Butterworths, pp 85–100

Clayton, C R I, Matthews, M C and Simons, N E (1995)
Site Investigation
2nd edn, Blackwell Science, Oxford

Corbett, B O (1982)
"The redevelopment of contaminated land"
The Structural Engineer, Vol 60A, No 1, January, pp 15–21

Cragg, C B H and Walker, B P (1987)
"To accept or avoid settlement when developing marginal land"
In: *Building on derelict and marginal land*, Proc Conf, Glasgow, 1986
Thomas Telford, London, pp 59–67

Crossley, A D and Thomson, G H (1987)
"Land development involving ground treatment by dynamic compaction"
In: *Building on derelict and marginal land*, Proc Conf, Glasgow, 1986
Thomas Telford, London, pp 785–790

Crowhurst, D and Manchester, S J (1993)
The measurement of methane and other gases from the ground
Report 131, CIRIA, London

Dale, A (1987)
"The role played by the Scottish Development Agency in the treatment of and building on marginal and derelict land. Keynote address"
In: *Building on derelict and marginal land*, Proc Conf, Glasgow, 1986
Thomas Telford, London

D'Appollonia, D A (1984)
"Contribution to discussion session: design and performance"
Proc 8th Eur Conf on Soil Mech and Found Engg, Helsinki, 1983
AA Balkema, Rotterdam, Vol 3, pp 1367–1369

Deacon, R C (1987)
Concrete ground floors: their design, construction and finish
Publication 48.034, Cement and Concrete Association, Wexham Springs

Dumbleton, M J (1980)
Historical investigations of site use in reclamation of contaminated land
Proc Conf Soc of Chemical Industry, London, pp B3/1–13

Dumbleton, M J (1983)
Air photographs for investigating natural changes, past use and present condition of engineering sites
Laboratory Report 1085, Transport and Road Research Laboratory, Crowthorne

Dumbleton, M J and West, G (1970)
Air photograph interpretation for road engineers
Laboratory Report LR369, DoE: Transport and Road Research Laboratory, Crowthorne

Dumbleton, M.J. and West, G (1976)
Preliminary sources of information for site investigation in Britain
Laboratory Report LR403 (rev edn 1976), DoE: Transport and Road Research Laboratory, Crowthorne

Eakin, W R G and Crowther, J (1985)
"Geotechnical problems on land reclamation sites"
Municipal Engineer, October, pp 233–245

Engelhardt, K and Golding, H C (1976)
"Field testing to evaluate stone column performance in a seismic area"
In: *Ground treatment by deep compaction*, Instn Civ Engrs, London, pp 61–69

Golder, H Q (1964)
"State of the art of floating foundations"
J Soil Mech and Found Div, Am Soc Civ Engrs, SM2, pp 81–88

Golder, H Q (1975)
"Floating foundations"
In: Winterkorn and Fang (eds), *Foundation engineering handbook*
Van Nostrand Reinhold, New York, pp 537–555

Greenwood, D A (1991)
Load tests on stone columns
In: M I Esrig and R C Bachus (eds), *Deep Foundations Improvements: Design, Construction and Testing*
Special Technical Publication 1089, Am Soc Test Mat, Philadelphia, PA

Greenwood, D A and Kirsch, K (1984)
"Specialist ground treatment by vibratory and dynamic methods"
Proc Int Conf Advances in piling and ground treatment
Thomas Telford, London, pp 17–45

Gutt, W, Nixon, P J, Smith, M A, Harrison, W M and Russell, D (1974)
A survey of the locations, disposal and prospective uses of the major industrial by-products and waste materials
Current Paper CP19/74, Building Research Establishment, Garston

Hartless, R (comp) (1992)
Methane and associated hazards to construction: a bibliography
Special Publication 79, CIRIA, London

Healy, P and Head, J M (1984)
Construction over abandoned mine workings
Special Publication 32, CIRIA, London

Heathcote, F W L (1965)
"Movement of articulated buildings on subsidence sites"
Proc Instn Civ Engrs, Vol 30, pp 347–368

Henkel, D J (1984)
"Design and work performance: British practice"
Proc 8th Eur Conf on Soil Mech and Found Engg, Helsinki
AA Balkema, Rotterdam, Vol 3

Hooker, P J and Bannon, M P (1993)
Methane: its occurrence and hazards in construction
Report 130, CIRIA, London

ICE (Institution of Civil Engineers) 1987
Specification for ground treatment
Thomas Telford, London

Leach, B A and Goodger, H K (1991)
Building on derelict land
Special Publication 78, CIRIA, London

McEntee, J M (1987)
Review paper in: *Building on marginal and derelict land*
Proc Int Conf, Thomas Telford, London, pp 381–389

Mitchell, J K (1982)
"Soil improvement: state-of-the-art report"
Proc 10th Int Conf on Soil Mech and Found Engg, Stockholm, Sweden, 1981
AA Balkema, Rotterdam, Vol 4, pp 509–565

Mitchell, J K (1986)
"Ground improvement evaluation by in-situ tests"
In: *Use of in-situ tests in geotechnical engineering*, Proc Specialty Conf, Virginia,
Geotechnical Special Publication No 6, Am Soc Civ Engrs, pp 221–236

Mitchell, J M (1987)
Discussion contrib in: *Building on marginal and derelict land*, Proc Conf, Glasgow, 1986
Thomas Telford, London, pp 581–594

National Economic Development Office (NEDO) (1985)
*Thinking about building: a successful business customer's guide to using the
construction industry*
National Economic Development Office, London

NEDO (1988)
Faster building for commerce
National Economic Development Office, London

National House-Building Council (NHBC) (1995)
Chapter 4.6 "Vibratory ground improvement techniques"
In: *NHBC Standards*, April, National House-Building Council, London

Neal, F R and Judge, C J (1987)
Classes of imposed floor loads for warehouses
BRE Information Paper IP 19/87, Building Research Establishment, Garston

Nicholson, D P, Tse, C-M, and Penny, C W (1999)
The Observational Method in ground engineering: principles and applications
Report 185, CIRIA, London

Seago, K and Treharne, G (1985)
Development on waste/derelict land: a designer's viewpoint
Internal report, Ove Arup and Partners, London

Skilton, E J (1987)
"Vibro-flotation and jackable foundations combine to cut costs"
In: *Building on marginal and derelict land*, Proc Conf, Glasgow, 1986
Thomas Telford, London, pp 759–764

Skipp, B O and Hall, M J (1982)
Health and safety aspects of grout treatment materials
Report 95, CIRIA, London

Slocombe, B C (1989)
"Thornton Road, Listerhills, Bradford"
In: *Piling and Deep Foundations*, Proc Int Conf, London, 1989
AA Balkema, Rotterdam, Vol 1, pp 131–142

Slocombe, B C and Moseley, M P (1987)
"Experience with dynamic compaction on derelict sites"
In: *Building on marginal and derelict land*, Proc Conf, Glasgow, 1986
Thomas Telford, London, pp 799–806

Somerville, M A (1987)
"Infilling a lock with colliery discard"
In: *Building on marginal and derelict land*, Proc Conf, Glasgow, 1986
Thomas Telford, London, pp 821–826

Swain, A and Holt, D N (1987)
"Dynamic compaction of a refuse site for a road interchange"
In: *Building on marginal and derelict land*, Proc Conf, Glasgow, 1986
Thomas Telford, London, pp 339–357

Thomas Telford Ltd (1987)
Building on marginal and derelict land
Proc Conf, Glasgow, 1986
Thomas Telford, London

Thomas Telford Ltd (1993)
Site Investigation Steering Group publications
1. *Without site investigation ground is a hazard*
2. *Planning, procurement and quality management*
3. *Specification for ground investigation*
Thomas Telford Limited, London

Tomlinson, M J and Wilson, S M (1973)
"Preloading of foundations by surcharge on filled ground"
Géotechnique, Vol 23, No 1, March, pp 117–120

Uff, J F and Clayton, C R I (1988)
Recommendations for the procurement of ground investigation
Special Publication 45, CIRIA/BRE, CIRIA, London

Weltman, A J (1983)
"Use of aerial infra-red photography for the detection of methane from landfills"
Ground Engineering, Vol 16, No 3, April, pp 22–23

Weltman, A J and Head, J M (1983)
Site investigation manual
Special Publication 25, CIRIA, London

Whitelaw, J (1986)
"Mini-plant flies infra-red sorties"
New Civil Engineer, No 685, 10 Apr, pp 39–40

4   Improvement by vibration

Vibration can be used to compact soils and fills. Vibrating rollers are familiar because of their use to compact relatively thin layers of earth fill and bituminous road materials. Using vibration to densify the ground in place and to depth involves either penetrating the ground with the vibrator or applying a very high level of energy at the ground surface or at single points within the ground.

The densification is achieved by a combination of ground displacement and vibration, in most cases with the addition of new material into the ground. Thus the insertion of a probe vibrator displaces the ground; the continuing vibration of the probe can compact the ground further; sand or stones vibrated into the ground remain as columns of dense material within it. Similarly a compaction pile driven into the ground remains there after first displacing the ground aside, densifying it further as the pile is driven down. With dynamic compaction, the first drop of the heavy weight (the pounder) displaces the ground downwards and sideways; subsequent drops (and those on adjacent areas) continue the densification, and selected fill is used above and in the craters left by the pounder to replace the ground volume change.

Techniques employing vibration in some manner are described in the following order:

- vibro-compaction – Section 4.1
- vibro stone columns – Section 4.2
- dynamic compaction (heavy tamping) – Section 4.3
- rapid impact compaction – Section 4.4
- vibratory probing (Vibro-wing, Terraprobe, Y-probe) – Section 4.5
- compaction piles – Section 4.6
- blasting – Section 4.7.

4.1   VIBRO-COMPACTION

4.1.1   Definition

Vibro-compaction (or vibro-flotation) is a technique for increasing the density of granular soils by the insertion to depth of a heavy vibrating poker, known as a depth vibrator, vibroflot or flot, which is inserted into the ground in conjunction with water flushing. The grading range of granular soils usually suitable for vibro-compaction is shown in Figure 4.1 (from Brown, 1977).

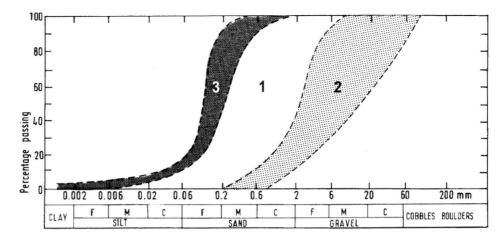

Grading zone 1: most suitable – very loose sands below the water table.

Grading zone 2: slower penetration rates or adverse situations of dense sands and gravels, cemented sands or deep water table.

Grading zone 3: more difficult to compact if soil gradings entirely within this zone. Achieved densification reduces with increasing silt and clay content. Clay and organic layers and cementing reduce effectiveness.

Figure 4.1   *Grading ranges of soils usually suitable for vibro-compaction (after Brown, 1977)*

4.1.2   Principle

The vibrating poker imparts essentially horizontal vibrations into the ground that overcome the frictional contacts or effective stress between soil particles. The ground then settles into a state of greater relative density (Figure 4.2). An idealised response to vibration is shown on Figure 4.3. This shows that the soil structure is destroyed under dynamic stresses of about 0.5 g. With greater accelerations the shear strength decreases until at about 1–1.5 g the ground becomes fluidised. Beyond about 3 g, dilation of the soil can occur.

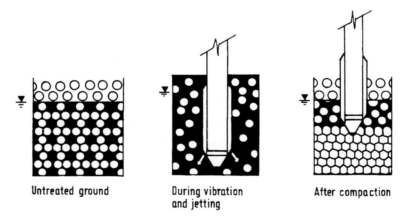

| Untreated ground | During vibration and jetting | After compaction |

Figure 4.2   *Principle of vibro-compaction (after Dobson, 1987)*

Accelerations imparted to the vibrator decrease with distance from it and Figure 4.3 also illustrates the several annular zones of soil response that can be identified. There is a lower state of compaction nearest the treatment poInt The key to successful compaction is that the ground can drain freely. In effect, therefore, permeabilities of 10^{-5} m/s or greater are necessary. The vibratory effects are also greatly affected by increasing silt content, as shear stresses are not transferred so effectively and the vibrations are severely damped. Figure 4.4 indicates how higher fines contents can reduce the degree of improvement that is achieved.

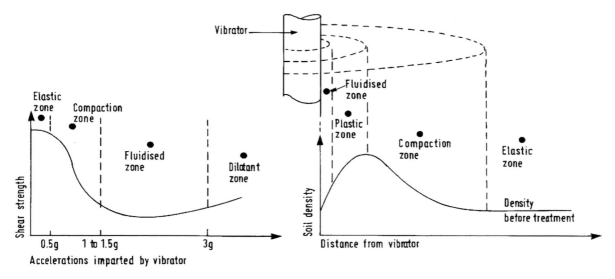

Figure 4.3   *Idealised response to vibration (after Greenwood and Kirsch, 1984)*

4.1.3   Description

The vibrating poker consists of the vibrating unit, follower sections and a lifting head. The source of the vibration is an eccentric weight mounted at the bottom of a shaft linked to a hydraulic or electric motor. This arrangement is shown in Figure 4.5. Greenwood and Kirsch (1984) suggest that power developed by the rigs ranges from about 35 kW to 100 kW, although some machines have a capacity of about 160 kW. Vibration frequencies are usually either 30 Hz or 50 Hz and give amplitudes of 5–10 mm, although there are depth vibrators with amplitudes up to 23 mm. Penetration during boring is assisted by the use of compressed air or water jets. Compaction or improvement of the ground takes place as the vibrating poker is raised in steps of about 0.3 m. To maximise the degree of compaction, sufficient time at each stage should be allowed at each step. Figure 4.6 illustrates the completion of one compaction poInt Treatment points are usually at spacings of 1.8–3.0 m, on either a triangular or square grid.

4.1.4   Applications

Vibro-compaction was first used for saturated natural fine sands, but it was soon applied to hydraulic fills (Jebe and Bartels, 1983). The process is often used to reduce the risk of liquefaction from a seismic event. Vibro-compaction has been used over water. The vibrators can be coupled together in groups of three or four. Depths of about 30 m can be treated or, occasionally, to about 50 m (Davis *et al*, 1981).

The geology of the UK seems rarely to allow the use of vibro-compaction. However, Bell *et al* (1987) report results of improving the density of hydraulic fill to depths of 28.5 m in conjunction with surcharging.

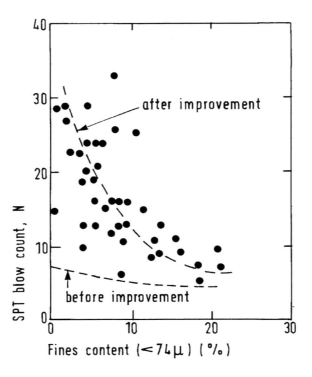

Figure 4.4   *Effect of increasing fines content on effectiveness of vibratory compaction (after Saito, 1977)*

4.1.5   Limitations

Vibro-compaction cannot be employed effectively where the fines content (silt and clay) exceeds about 20 per cent. This is illustrated by Figures 4.4 and 4.7, which also suggests the coarser grain size limit of applicability of vibro-compaction.

4.1.6   Design

Spacing of treatment points can be determined using empirical graphs of influence coefficients. Glover (1985) has summarised the work of D'Appollonia *et al* (1955), Webb and Hall (1969) and Brown (1977). The vibro-compaction treatment should extend outside the intended area of loading.

4.1.7   Controls

In the granular soils for which vibro-compaction is a suitable treatment, it is not possible to determine actual or relative densities directly – except by very sophisticated techniques for "undisturbed" sampling. Cone penetration testing or SPTs can be used as indicators of density increases, ie from comparative tests before and after treatment. Specifying required penetration resistances (or the implied relative density equivalent to, say, an SPT blow-count) should be based on site-specific field trials or an appropriate well-established published relationship. Controls also include spacing, depth, vibration energy and duration.

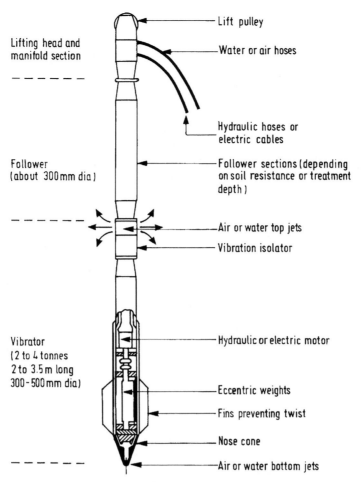

Figure 4.5 *Typical vibrator for vibro-replacement and vibro-compaction (after Greenwood, 1970)*

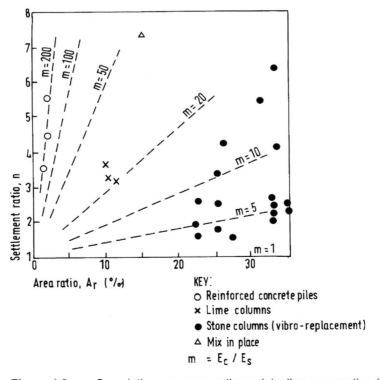

Figure 4.6 *Completing one compaction point: vibro-compaction (after Brown, 1977)*

4.1.8   References

Bell, A L, Slocombe, B C, Nesbit, A M and Finey, J T (1987)
"Vibro-compaction densification of a deep hydraulic fill"
In: *Building on marginal and derelict land*, Proc Conf, Glasgow, 1986
Thomas Telford, London, pp 697–712

Brown, R E (1977)
"Vibroflotation compaction of cohesionless soils"
J Geotech Engg Div, Am Soc Civ Engrs, Vol 103, GT2, Dec, pp 1437–1451

D'Appollonia, E, Miller, C E and Ware, T M (1955)
"Sand compaction by vibroflotation"
Trans Am Soc Civ Engrs, Vol 120m, pp 154–168

Davis, P, Neilssen, H and Pladet, A (1981)
"Mytilus – a soil compaction vessel"
Proc 10th Int Conf Soil Mech and Found Engg, Stockholm
AA Balkema, Rotterdam, Vol 3, pp 641–644

Dobson, T (1987)
"Case histories of the vibro system to minimise the risk of liquefaction"
In: *Soil improvement – a ten year update*
Geotech Spec Pub 12, Am Soc Civ Engrs, pp 215–231

Greenwood, D A (1970)
Mechanical improvement of soils below ground surface
In: *Ground Engineering*, Proc Conf, Instn Civ Engrs, London, pp 11–22

Greenwood, D A and Kirsch, K (1984)
"Specialist ground treatment by vibratory and dynamic methods"
In: *Advances in piling and ground treatment*, Proc Int Conf, London
Thomas Telford, London, pp 17–45

Glover, J C (1985)
"Sand compaction and stone columns by the vibroflotation process"
In: *Recent developments in ground improvement techniques*, Proc Int Symp, Bangkok
AA Balkema, Rotterdam, pp 3–15

Jebe, W and Bartels, K (1983)
"The development of compaction methods with vibrators from 1976 to 1982"
Proc 8th Eur Conf Soil Mech and Found Engg, Helsinki, 1983
AA Balkema, Rotterdam, Vol 1, pp 259–266

Saito, A (1977)
"Characteristics of penetration resistance of a reclaimed sandy deposit and their changes through vibratory compaction"
Soils and Foundations, Vol 17, No 4, December, pp 31–43

Webb, D L and Hall, R I (1969)
"Effects of vibroflotation on clayey sands"
J Soil Mech and Found Div, Am Soc Civ Engrs, Vol 95, SM6, November, pp 1365–1378

4.2 VIBRO STONE COLUMNS

4.2.1 Definition

The formation of columns usually of compacted stone in the ground using a heavy vibrating poker to displace the *in-situ* ground and to compact the imported material is referred to as vibro stone columns or vibro-replacement (Watts, 2000). The method is usually applied to made or filled ground and cohesive soils; the wide range of its applicability is indicated on Figure 4.7. The vibratory poker is also called to as a vibroflot, flot or depth vibrator.

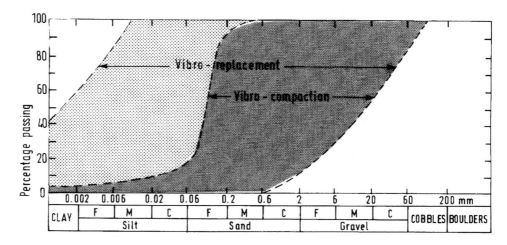

Figure 4.7   *Grading range of soils usually suitable for vibro-compaction and vibro-replacement*

4.2.2 Principle

The principle is illustrated in Figure 4.8, showing the replacement of loose soil by compacted stone, possibly with densification of the surrounding granular material. Introducing columns of stone within the ground stiffens the soil mass and the columns themselves are stiffer than the surrounding ground (see Figure 2.1). The stone columns therefore carry most of the imposed load at first, but under continuing load, they try to expand, or bulge, into the ground mass. The capacity of the stone column then depends on the degree of the constraint offered by the ground itself.

4.2.3 Description

The stone column methods were developed independently in Germany and the UK in the late 1950s. The basic equipment of vibrator and track-mounted rig remains essentially the same, but there are now improved systems of flushing, backfill handling, vibration and construction control. An extension of the technique is the formation of vibro concrete columns (VCC) placed by similar means in the ground. This method is increasingly offered and used, but there are few references reporting its performance. As it is rather like foundation piling, the method is not discussed further here, other than to note that the effect on the soil between columns will be similar to that with stone columns during installation, but there should be greater rigidity in the concrete column and it will not be so permeable.

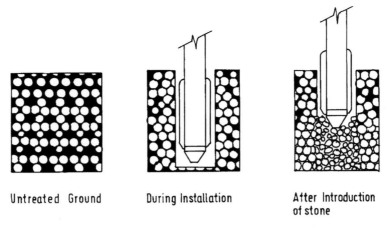

Untreated Ground   During Installation   After Introduction
                                           of stone

Figure 4.8   *Principle of vibro-replacement (after Dobson, 1987)*

Figure 4.5 shows a typical vibrator, which is powered either hydraulically or electrically (also see Section 4.1.3). Figure 4.9 is an example of present-day equipment in operation.

When forming a stone column by vibro-replacement, the vibrator and extension tubes, supported by a suitable base machine, are placed over the selected position (Figure 4.10). With the motor switched on, the vibrator is lowered into the ground until the desired depth is reached. Water or air jetting may be used to assist penetration of the vibrator, as can a pull-down facility, available on purpose-built base machines with vibrators mounted on leaders.

After penetration, the vibrator is partly or fully withdrawn and the hole created is filled with a charge of crushed rock aggregate (the stone), typically enough to fill about 1 m of the hole. The vibrator is lowered into the stone and with its specially designed nose cone it displaces the stone both radially and downwards into the surrounding soil. This displacement of the soil is therefore additional to that created by the initial penetration of the vibrator. By further cycles of this process, a compact column of stone interlocked with the surrounding ground is built up to the ground surface.

In this process, stone reaches the base of the hole formed by the vibrator either through the annulus between the vibrator and the surrounding soil or directly down the hole when the vibrator is entirely withdrawn. This method is referred to as the top-feed system and can be applied dry when the bore is stable on removal of the depth vibrator. In unstable soils, the stone columns may end up with inclusions of soil in them brought down from the side of the hole. The wet top-feed system or dry bottom-feed methods are for unstable soils, normally below the water table, where the depth vibrator is only partly withdrawn to charge the ground with stone. In the bottom-feed system the stone is fed to the nose of the vibrator through pipes attached to the vibrator and extension tubes. The stone is placed at the point of vibration and compaction, producing a more consistent stone column.

Figure 4.9 *Vibro-replacement plant*

Columns are typically 0.6–0.8 m in diameter with near-parallel or sometimes tapered sides. Typical spacings are 1–2 m for isolated loads, and 2–3 m for widespread loads such as ground slabs. Grids can be triangular or square. Depths of greater than 30 m can be treated. Stone columns in filled ground are shown exposed by excavation in Figure 4.11 (Mitchell, 1987).

The method statement that should be provided about the equipment and method should include the following information:

- type and energy rating of vibratory probe
- number of insertions of the vibratory probe
- expected noise and vibration levels
- details/records of power output and depth of penetration
- backfill material to be used.

Figure 4.10 *Vibro-replacement: construction process*

Figure 4.11 *Stone column for strip footing of three-storey housing project; demolition debris surrounding the column*

4.2.4 Applications

Vibro-replacement is mainly used for building and light industrial development projects, ie for structures not particularly sensitive to settlement deformations. Warehouses, oil tanks, buildings of up to four storeys and, less often, rigid multi-storey buildings have been built on ground improved by vibro-replacement. Typically, strip footings are placed on the treated ground, after the upper 0.6 m of the columns and surrounding soil has been trimmed and replaced with compacted granular fill – the base of the footing being about 1 m below the final ground level (see Figure 4.11). Note that changes in the site ground level or surcharges affect the behaviour of the improved ground.

Increasingly, the vibro stone column method is being used to reduce the liquefaction potential of loose fine sands in layered deposits. In this case, three effects contribute to the improvement: the introduction of stronger material, the provision of drains to dissipate or prevent the build-up of high pore pressures, and the increase in the density of the *in-situ* soil. Kirsch and Chambosse (1981), Bhandari (1984), and Mitchell and Huber (1985) provide examples. The method is also used to improve the foundation stability of embankments on soft soils. Here, as well as increasing the shearing resistance by the inclusion of stronger material, the stone columns may also act as vertical drains, speeding up consolidation of the soft soil under the embankment loading.

By reducing the variability in compressibility of the ground mass, vibro-replacement controls settlement. It is therefore also used below embankments. Examples are given by Vautrain (1978), Di Maggio and Goughnour (1979), and Munfakh *et al* (1984).

4.2.5 Limitations

Table 4.1 lists the limitations of vibro stone column treatments. These situations do not necessarily preclude the use of stone columns, but they may indicate that the approach to the work should be modified. Note that the use of vibro concrete columns can overcome some of the limitations of lack of lateral support.

4.2.6 Design

The behaviour of a stone column within ground is difficult to model mathematically. A review of the various methods is given in Greenwood and Kirsch (1984), probably still the best current guide to design. The book discusses the ultimate capacity of isolated columns, their settlement, and the settlement of widespread loading on column groups. In general, the method of Priebe (1976) was used for large groups in cohesionless ground, and that of Hughes and Withers (1974) to determine bearing capacity in cohesive ground. More recently, Priebe (1995) has updated the guidance on design and extended it to cover design of vibro-replacement treatments to minimise liquefaction risks (Priebe, 1998). Settlements are assessed using conventional soil mechanics methods. The settlement of ground improved by vibro-replacement can typically be about half that of the untreated ground.

As the columns are stiffer than the surrounding ground (which may itself be little changed), floor slabs will not necessarily be supported across their full area, but more or less direct on the columns. Thus the design of the slabs should explicitly consider this possibility.

Table 4.1 *Adverse situations for vibro-replacement*

Adverse situation	Possible difficulty
Very soft clay (s_u<15 kN/m²) sensitive clays, silts and silty fine sands	Lack of lateral support, contamination
Peat thicker than about 0.6 m or in several layers	Lack of lateral support
Obstructions, hard ground	Incomplete penetration
Filled ground with voids and/or obstructions	Incomplete penetration and large subsequent settlement
Loose fills susceptible to collapse settlement on inundation or rises in groundwater	Sudden loss of lateral support at top of column when fill settles; stone columns act as drains, allowing water to enter the ground
Fills still settling under self-weight	Shear transmitted to stone column, lateral support at top reducing with time
Contaminated and methane-generating ground	Stone columns act as collectors of leachate and vents for gas
Edges of filled pits or quarries or discontinuity in geometry of sound underlying strata	Incompatible compressibilities across proposed foundation
Near crest or toe of slope	Reduction in slope stability
Nearby trees	Roots may impede formation of stone columns and, subsequently, grow through and in the columns

4.2.7 Controls

Vibro-replacement is primarily undertaken to manage the potential for settlement. Large-scale loading tests are thus the best indicator of the settlement characteristics of the treated ground under load. But as these are costly and time-consuming, they cannot be used as controls of workmanship. Rather this is done by observation and recording, now usually electronically, of the basic parts of the process, ie positions, depths, quantities, feed rates, withdrawal and compaction times etc. See also the BRE specification (Watts, 2000).

4.2.8 References

Bhandari, R K M (1984)
"Behaviour of a tank founded on soil reinforced with stone columns"
Proc 8th Eur Conf Soil Mech and Found Engg, Helsinki, 1983
AA Balkema, Rotterdam, Vol 1, pp 209–12

Di Maggio, J A and Goughnour, R D (1979)
"Demonstration program on stone columns"
In: *Soil reinforcement*, Proc Int Conf, Paris
Editions Anciens, Ecole Nationale des Ponts et Chaussées, pp 249–254

Dobson, T (1987)
"Case histories of the vibro systems to minimise the risk of liquefaction"
In: *Soil improvement – a ten year update*
Geotech Spec Pub 12, Am Soc Civ Engrs, pp 215–223

Greenwood, D A and Kirsch, K (1984)
"Specialist ground treatment by vibratory and dynamic measures"
In: *Advances in piling and ground treatment*, Proc Int Conf, London
Thomas Telford, London, pp 17–45

Hughes, J M and Withers, N J (1974)
"Reinforcing of soft cohesive soils with stone columns"
Ground Engineering, Vol 7, No 3, May, pp 42–49

Kirsch, K and Chambosse, G (1981)
"Deep vibratory compaction provides foundations for two major overseas projects"
Ground Engineering, Vol 14, No 8, November, pp 31–38

Mitchell, J M (1987)
Contribution to discussion
In: *Building on marginal and derelict land*, Proc Conf, Glasgow, 1986
Thomas Telford, London, pp 581–594

Mitchell, J K and Huber, T R (1985)
"Performance of a stone column foundation"
J Geotech Engg, Am Soc Civ Engrs, Vol 111, No 2, February, pp 205–223

Munfakh, G A, Sarkar, S K and Castelli, R J (1984)
"Performance of a test embankment founded on stone columns"
In: *Piling and ground treatment*, Thomas Telford, London, pp 259–265

Priebe, H-J (1976)
"Evaluation of the settlement reduction of a foundation improved by vibro-replacement"
Bautechnik, Vol 5, pp 160–162 (in German)

Priebe, H-J (1995)
"The design of vibro-replacement"
Ground Engineering, Vol 28, No 10

Priebe, H-J (1998)
"Vibro-replacement to prevent earthquake induced liquefaction"
Ground Engineering, Vol 31, No 9, September, pp 30–33

Vautrain, J (1978)
"Reinforced earth wall on stone columns in soil"
Bulletin de Liaison, Spec Issue VIE, Lab des Ponts et Chaussées, Paris, pp 188–94

Watts, K S (2000)
Specifying vibro stone columns
Publication BR391, Building Research Establishment, CRC, Watford

4.3 DYNAMIC COMPACTION

4.3.1 Definition

Dynamic compaction is the process of systematically tamping the ground surface with a heavy weight dropped from height. It is used to improve the bearing capacity of a wide range of materials, generally loose fills in the UK. The process is sometimes referred to as heavy tamping or pounding, dynamic pre-compression or dynamic consolidation.

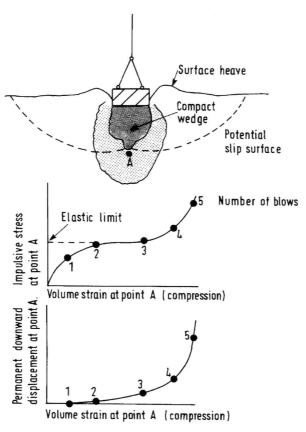

Figure 4.12 *Idealised stress and displacement at a point below dynamic compaction (after Greenwood and Kirsch, 1984)*

4.3.2 Principle

Greenwood and Kirsch (1984) describe how the tamping induces both punching shear and compaction displacement, as the momentum of the falling weight decays. Figure 4.12 illustrates the idealised stress and displacement at a point within the compaction zone. It can be seen that there is no further benefit from continued tamping after voids in the ground become closed, ie at point A the ground does not compress further. Even though point A may be driven further down, this is because surrounding soil is being displaced to the side and upwards. Excess pore-water pressures must also be allowed to dissipate or else they will substantially inhibit the effectiveness of the compaction process. Therefore, saturated soils have to be and remain free-draining for the process to work. Many fills, however, are dry with no pore-water pressures to inhibit use of the method. Figure 4.13 is the result of a laboratory model test on sand, which illustrates how the zone of compacted material grows. Figure 4.14 shows a cut face through soil that had been dynamically compacted.

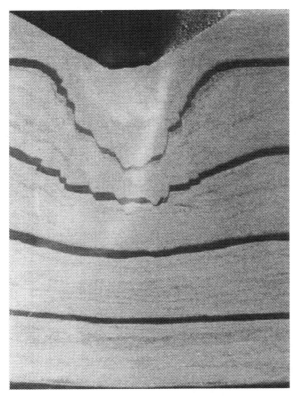

Figure 4.13 *Laboratory modelling of dynamic compaction (6.7 kg tamper, 1.5 m drop). The deformation and shear displacements after 15 drops are marked*

Figure 4.14 *Exposed face of dynamically compacted ground*

4.3.3 Description

Typically, weights of 5–20 tonnes are used with drops of up to about 20 m. Weights are often mass concrete or steel plates welded together and are usually about 2 m square. A layer of granular fill, typically 0.3–1.0 m thick, is placed over the site first. The weight is dropped on a square grid of about 5–10 m. Five to ten blows are applied to each imprint in each pass of the weight (Greenwood and Kirsch, 1984). The aim is to produce a crater about 0.5–2 m deep. Groundwater may have to be lowered to ensure that it does not enter the imprints. These are backfilled after each pass of the weight. The process is repeated until the required induced settlement of the site is achieved. Depending on the ground and the type of improvement that is needed, induced settlements can range from 0.5 m to 2 m. Alternatively, the process may be controlled by achieving specified minimum or average *in-situ* properties determined by penetration tests, either SPT or CPT. Figure 4.15 shows a typical site being improved by dynamic compaction.

Figure 4.15 *General view of first pass of dynamic compaction at a site in Cwmbran*

4.3.4 Applications

Dynamic compaction can be applied to improve ground to depths to 15 m or so, usually on sites of 5000 m^2 or greater area. The process has been used mainly to compact sands, silty sands, hydraulic fill and silty clay fills. Mayne *et al* (1984), in their collection of data from 124 sites, reported the increasing use of dynamic compaction to improve rubble, rockfill, domestic refuse and colliery-waste fills. Welsh (1983) reports how treatment can cause up to 25 per cent compression of 10 m of refuse. In addition, dynamic compaction has been applied to karstic conditions in an attempt to create sinkholes before construction (Wagener, 1984; Guyot, 1984). Further case histories can be found in the Glasgow Conference on Marginal and Derelict Land (Thomas Telford, 1987).

4.3.5 Limitations

Dynamic compaction is not effective in soft alluvial or marine clays (Choa *et al*, 1979; Charles and Watts, 1982). Harder layers within the ground being improved can also inhibit the effectiveness of the treatment. Further limitations or conditions to be considered are listed in Table 5.2. Particular attention is drawn to the possible need for dewatering prior to treatment. The effects of ground-borne vibration also have to be considered (Greenwood and Kirsch, 1984; Head and Jardine, 1992). Figure 4.16 shows typical measured vibrations reported by Greenwood and Kirsch (1984). It is probably prudent to limit ground improvement by dynamic compaction to not closer than about 30 m from a structure.

Table 4.2   *Adverse situations for dynamic compaction*

Adverse situation	Possible difficulty
Soft clays ($s_u < 30$ kN/m^2)	Insufficient resistance to transmit tamper impulse
High groundwater level	Need to dewater and to consider possible effects of subsequent recovery in water level
Vibration effects (may be worse if groundwater level is high)	Distance from closest structure to be of the order of 30 m or more
Clay surface	May be inadequate for heavy cranes and unsuitable for imprint backfilling
Clay fills	May be subject to collapse settlement if inundated later
Flying debris	Precautions for site and public safety
Voided ground or Karst features below treated ground	Treatment may not reach the voided zone or may make it less stable
Biologically degrading material	Compaction may create anaerobic conditions and regenerate or change the seat of the biological degradation

4.3.6   Design

Mayne *et al* (1984) related apparent maximum discernible depth of improvement to energy per blow for 110 sites. Figure 4.17 shows the data that indicate the depth of influence $D_{max} \cong n (WH)^{1/2}$ where, as a first approximation, $n = 0.5$, and W is the mass and H is the drop. Treatment should also be extended to at least the distance of the compressible layer, or depth of compaction, outside the building area. Comprehensive design and construction guidelines for dynamic compaction for highway construction are given in a report of the US Federal Highway Administration (Lukas, 1986).

4.3.6   Controls

The basic controls on workmanship are in positioning and careful dropping of the tamper. Checks should be made to note and record the energy (number of drops) imparted at each imprint and the quantity of imported fill to make up the general site level. By its nature, the process has to be linked to set procedures to keep within environmental constraints, such as vibration and noise limits. Vibration or noise monitoring may be necessary, which should also be used to establish background levels.

Indicators of the improvement achieved might be in terms of:

- a depth of lowering of the site surface (or the volume required to maintain the original level)
- comparative changes in penetration test resistances
- large-scale (zone or skip) loading tests.

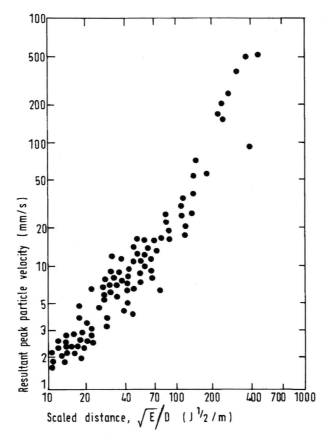

Notes

1. Resultant peak particle velocities are derived from independent measurements of horizontal, vertical and shear waves.

2. *E* is energy in one drop in Joules. Above observations are on tamping using drop energies of from 1000 to 3200 kJ, ie of 10 t × 10 m; 10 t × 20 m; 15 t × 16 m; 16 t × 20 m.

3. *D* is distance in metres.

Figure 4.16 *Observations of resultant peak particle velocities with distance from tamping point (after Greenwood and Kirsch, 1984)*

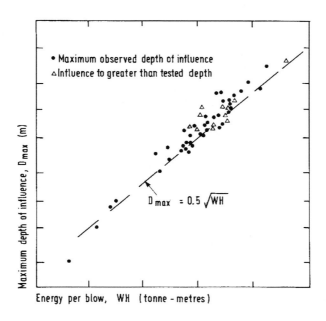

Figure 4.17 *Depth of influence and energy per blow for dynamic compaction (after Mayne et al, 1984)*

4.3.7 References

Charles, J A and Watts, K S (1982)
"A field study of the use of Dynamic Consolidation ground treatment technique on soft alluvial soil"
Ground Engineering, Vol 15, No 5, July, pp 17–25

Choa, V, Vijiratam, A, Karunaratne, G P, Ramaswamy, S D and Lee, S L (1979)
"Consolidation of Changi Marine Clay of Singapore using flexible drains"
Proc 7th Eur Conf Soil Mech and Found Engg, Brighton
British Geotech Soc, London, Vol 3, pp 29–36

Greenwood, D A and Kirsch, K (1984)
"Specialist ground treatment by vibratory and dynamic methods"
In: *Advances in piling and ground treatment*, Proc Int Conf, London
Thomas Telford, London, pp 17–45

Guyot, C A (1984)
"Collapse and compaction of sinkholes by dynamic compaction"
In: *Sinkholes: their geology, engineering and environmental impact*,
Proc Conf, Orlando, Florida
AA Balkema, Rotterdam, pp 419–23

Head, J M and Jardine, F M (1992)
Ground-borne vibrations arising from piling
Technical Note 142, CIRIA, London

Lukas, R G (1986)
Dynamic compaction for highway construction, Vol I Design and construction guidelines
Final Report FHWA/RD-86/133, US Dept Transportation, Federal Highways Administration, Washington DC, p 230

Mayne, P W, Jones, J S and Dumas, J C (1984)
"Ground response to dynamic compaction"
J Geotech Engg Div, Am Soc Civ Engrs, Vol 110, No 6, June

Thomas Telford Limited (1987)
Building on marginal and derelict land, Proc Conf, Glasgow, 1986
Thomas Telford, London

Wagener, F von M (1984)
"Engineering construction on dolomite"
Geotech Div, S African Instn Civ Engrs, Marshalltown, S Africa, pp 197–199

Welsh, J P (1983)
"Dynamic deep compaction of sanitary landfill to support superhighway"
Proc 8th Eur Conf Soil Mech and Found Engg, Helsinki, Vol 1
AA Balkema, Rotterdam, pp 319–321

4.4 RAPID IMPACT COMPACTION

4.4.1 Definition

Rapid impact compaction is the application of the energy from repeated blows of a relatively high frequency hydraulic hammer to the ground through an anvil in a tamping foot resting directly on the ground to be compacted. The technique is also called rapid dynamic compaction. It was developed by the Ministry of Defence for the rapid repair of bomb craters in airfield runways, but was later extended to commercial operation. The hammer and anvil are mounted on a crawler rig for site mobility (Figure 4.18).

4.4.2 Principle

The principle of imparting energy from a drop weight is the same as for dynamic compaction, but the technique differs in several respects. First, the energy of each drop is very much less than with dynamic compaction. Second, the frequency of the impacts is of the order of 40–60 per minute. A third difference is that the drop hammer impacts on an anvil built into the tamping foot rather than on to the ground surface as with dynamic compaction.

Figure 4.18   *Rapid impact compactor*

4.4.3 Description

The specially designed BSP hydraulic hammer has a mass of 7 t and a drop height of 1.2 m. Typical amounts of energy are 150–250 t-m/m^2 and at any one point up to 100 blows may be applied. The crawler rig weighs 50 t. Two types of tamping foot are available, one of 1.5 m diameter (Figure 4.19) and the other 1.8 m square.

A gravel layer is placed over the treatment area where the existing surface is poor. At the end of treatment the surface can be regraded and either tamped with the square plate or compacted by roller.

4.4.4 Applications and limitations

The technique can be used for similar situations to those where dynamic compaction would be appropriate, ie loose granular soils, loose fills, and landfill materials. It is not suitable for natural silts and clays. Depths of treatment of 3 to 4 m are typical but treatment to 10 m is claimed. An assessment of the system has been carried out by BRE (Watts and Charles, 1993).

Advantages of this method are the mobility of the rig, its potential for working reasonably close to existing structures, and the low risk of flying debris or injury because the tamper is in contact with the ground.

The effects of noise and vibration may have to be considered.

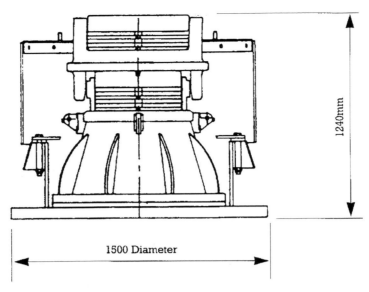

1240mm

1500 Diameter

Figure 4.19   *Tamping foot of rapid impact compactor*

4.4.5   Design

The tamping points are usually on a square or triangular grid initially at about 2 m spacings. Secondary tamping may be at intermediate positions. The treatment energy is judged by trials on site to find out when further blows produce little further settlement. A typical limit would be when the settlement per blow becomes less than 5 to 10 mm.

4.4.6   Controls

The basic controls on workmanship are in positioning and applying sufficient impact energy. Checks should be made to note and record the number of impacts imparted at each imprint and the quantity of imported fill to make up the general site level. By its nature, the process has to be linked to set procedures to keep within environmental constraints, such as vibration and noise limits. Thus vibration or noise monitoring may be necessary – which should also be used to establish background levels.

Indicators of the improvement achieved might be in terms of:

1   A depth of lowering of the site surface (or the volume required to maintain the original level).

2   Comparative changes in penetration test resistances.

3   Large-scale (zone or skip) loading tests.

4.4.7   Reference

Watts, K S and Charles, J A (1993)
"Initial assessment of a new rapid ground compactor"
In: *Engineered Fills* (B G Clarke, C J F P Jones and A I B Moffat, eds)
Thomas Telford, London, pp 349–412

4.5   VIBRATORY PROBING

4.5.1   Definition

Vibratory probing is a variation of vibro-compaction (Section 4.2) where granular soils are densified by probes that are vibrated at their head (ie above ground). Variously shaped probes have been used, of which four are described: the Terraprobe, the Y-probe, the Vibro-wing and the vibro-compozer. They are not in use in the UK.

4.5.2   Principle

The vibratory probe acts much in the same way as the poker or vibroflot described in Section 4.1, in that it imparts vibrations to depth in the ground, although these are likely to be predominantly vertical rather than horizontal. Figure 4.20 illustrates the process. The vibrations overcome the frictional contacts or effective stress between soil particles. In three of the four forms of vibratory probing described below no additional sand backfill is placed. However, in the fourth system (the vibro-compozer), sand backfill is compacted in the ground.

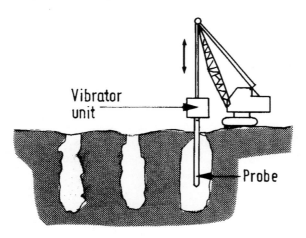

Figure 4.20   *Vibratory probing*

4.5.3   Description

Terraprobe

This device was developed in the United States and consists of a vibratory pile driver and probe. It is described by Anderson (1974). The probe consists of an open-ended casing made of 760 mm-diameter pipe with 10 mm wall thickness. It is attached to the vibrator with a hydraulic clamp to allow full transmission of the vibratory energy. Casing lengths are externally coupled to a length of between 14 m and 17 m. The casings are shown on Figure 4.21a and are usually vibrated in a square pattern at between 1 m and 2.5 m centres.

Y-Probe

The Y-Probe was developed in Belgium as an alternative to the Terraprobe. Wallays (1985) describes how the three-branch shape was chosen to eliminate the possibility of plugging (Figure 4.21b).

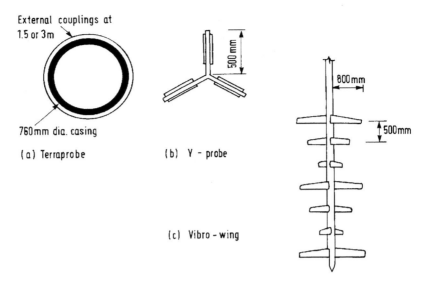

Figure 4.21 *Three types of vibratory probe*

The working principle is exactly the same as the Terraprobe. The range of soils suitable for its application is similar to that for vibro-compaction indicated on Figure 4.1, ie generally saturated fine sands. Wallays (1985) notes that the Y-Probe should not be used where the silt content exceeds 12 per cent. The maximum depth in which significant densification occurs appears to be about 10 m.

Vibro-wing

The Vibro-wing was developed in Sweden for compacting loose silty sands. It is described by Massarsch and Broms (1984) and Broms and Hansson (1984). Figure 4.21c shows the Vibro-wing equipment. It consists of a 15 m long steel rod, which is provided with 0.8 m long wings spaced at about 0.5 m centres. A heavy vibrator with a 70 kN mass is used to drive the rod into the ground quickly, taking perhaps one minute. Withdrawal is carried out more slowly, taking about five minutes. Probe spacings are typically on a 2.5 m triangular grid. As with other forms of vibro-compaction, silt content controls the effectiveness of this method.

A further development combines carefully designed perforations in a vertical probe comprising a steel column with a 2 m web and 1.5 m flanges. Used in combination with a variable-frequency vibrator to induce soil resonance, this allows wide spacing and is rapid, but is again limited by the silt/clay content (Massarsch, 1991).

Vibro-compozer

This method of vibro-compaction was developed for loose sands in Japan in about 1955, according to Aboshi *et al* (1979). Figure 4.22 illustrates the method of installation. A thick-wall casing is used, 0.4–1.5 m in diameter. Nakayama *et al* (1973) describe two types of vibrator, used at frequencies of 600–700 rpm. The standard and large models, with vibration forces of 20 t and 40 t, respectively, have weights of 4.5 t and 6.0 t and working vibrational amplitudes of 15–25 mm. Vibration forces of 60 t are available.

The casing is driven to the required depth under vibration, when it is filled with clean, coarse sand. By raising the casing a little, compressed air forces sand past the clack (one-way) valve at the base of the casing to start the sand column. The casing is then driven back to compact and expand the sand column.

The compacted sand column is usually about 0.7–2.0 m in diameter depending on the casing size. While the column is being formed, the sand level in the casing, depth of casing, weight of sand, and power consumption are monitored continuously.

The process is similar to the formation of Franki-type piles or compaction piles (Section 4.6) and could be classified as such. It is included in vibratory probing, however, because it employs a vibrator rather than a hammer.

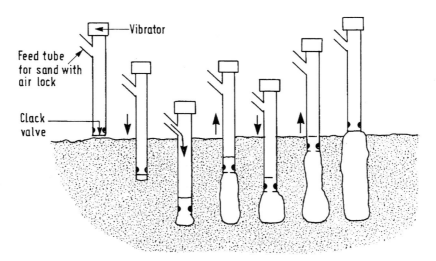

Figure 4.22   *Vibro-compozer process*

4.5.4   **Applications and limitations**

The Terraprobe, Y-Probe and Vibro-wing processes are not likely to be effective if there is more than about 10 per cent of silt and clay present. However, the Vibro-compozer method can be applied to soft clays as well as granular deposits. The sand columns tend to act as deep drains. Mizuno *et al* (1990) describe a recent trial application in Malaysian marine clays.

Janes (1973), Anderson (1974) and Leycure and Schroeder (1987) describe case histories of the use of the Terraprobe. The last case history compares the results of the method both on level ground and on a slope of hydraulic fill. Brown and Glenn (1976) and Faraco (1981) compare the Terraprobe with other methods of vibro-compaction.

Case histories of the use of the Y-Probe are reported by De Wolf *et al* (1983) and of casing driving by Wallays (1985). Wallays found that plugging of the casing occurred when the height of the sand column was about 10 times the inner diameter of the casing. This column of sand was very loose on withdrawal of the casing. In unsaturated ground, it was also necessary to decrease grid spacings to overcome capillary tensions in the ground. The method was compared with the poker or vibroflot method. For the Vibro-wing, only one large contract has been described in the literature, by Broms and Hansson (1984). Murayama and Ichimoto (1985) also compared the Vibro-compozer method with vibro-compaction carried out using a poker.

4.5.5 Design

Design using the Terraprobe, Y-Probe, and Vibro-wing is carried out best by reference to published case histories and the use of before-and-after *in-situ* testing such as SPT and CPT comparisons.

For the Vibro-compozer method in sands, Murayama and Ichimoto (1985) have published guidance for the spacing of the sand columns, based on achievable relative density and SPTs (Gibbs and Holtz, 1957). For clays, design methods are described by Aboshi *et al* (1979) and Murayama and Ichimoto (1982). Mitchell (1982) suggests that the design method formulated for use of sand columns to stabilise embankments can also be applied in the case of vibro stone columns (Section 4.3).

4.5.6 Controls

Controls are likely to be similar in nature to those of other forms of deep compaction, ie spacing and depth of treatment, duration and energy of vibration, volume and quality of any introduced aggregate material. Checks on effectiveness would be by *in-situ* testing such as CPT or SPT before and after treatment.

4.5.7 References

Aboshi, J, Ichomoto, E, Harada, K and Enoki, M (1979)
"The Compozer-method to improve characteristics of soft clays by inclusion of large diameter sand columns"
In: *Reinforcement of soils*, Proc Int Conf, Paris
Editions Anciens, Ecole Nationale des Ponts et Chaussées, Paris, pp 211–216

Anderson, R D (1974)
"New method for deep sand vibratory compaction"
Proc Am Soc Civ Engrs, J Const Div, Vol 100, CO1, March, pp 79–95

Broms, B B and Hansson, O (1984)
"Deep compaction with the vibro-wing method"
Ground Engineering, Vol 17, No 5, July, pp 34–36

Brown, R G and Glenn, A J (1976)
"Vibroflotation and Terra-Probe comparison"
Proc Am Soc Civ Engrs, J Geotech Div, Vol 102, GT10, October, pp 1059–1072

De Wolf, P., Carpentier, R., Allaert, J and De Rouck, J (1983)
"Ground improvement for the construction of the new outer harbour at Zeebrugge, Belgium"
Proc 8th Eur Conf Soil Mech and Found Engg, Helsinki, 1983
AA Balkema, Rotterdam, Vol 2, pp 827–832

Faraco, C (1981)
"Deep compaction field tests in Puerto de la Luz"
Proc 10th Int Conf Soil Mech and Found Engg, Stockholm, 1981
AA Balkema, Rotterdam, Vol 3, pp 659–662

Gibbs, H J and Holtz, W G (1957)
"Research in determining the density of sands by spoon penetration testing"
Proc 4th Int Conf Soil Mech and Found Engg, London

Janes, H W (1973)
"Densification of sand for drydock by Terra-Probe"
Proc Am Soc Civ Engrs, J Soil Mech and Founds Div, Vol 99, SM6, June, pp 451–420

Leycure, P and Schroeder, W L (1987)
"Slope effects on probe densification of sands"
In: *Soil improvement – a ten year update* (J P Welsh, ed), Proc Symp Geotech
Special Publication 12, Am Soc Civ Engrs, pp 197–214

Massarsch, K R (1991)
"Deep soil compaction using vibratory probes"
In: *Deep Foundations Improvements: Design, Construction and Testing* (M I Esrig and
R C Bachus, eds)
Special Technical Publication 1089, Am SocTest Mat, Philadelphia, PA

Massarsch, K R and Broms, B B (1984)
"Soil compaction by vibro-wing method"
Proc 8th Eur Conf on Soil Mech and Found Engg, Helsinki, 1983
AA Balkema, Rotterdam, Vol 1, pp 275–278

Mitchell, J K (1982)
"Soil improvement – state-of-the-art"
Proc 10th Int Conf Soil Mech and Found Engg, Stockholm, 1981
AA Balkema, Rotterdam, Vol 4, pp 509–565

Mizuno, Y, Shibata, W and Kanda, Y (1990)
"Trial embankments with sand compaction pile method at Muar Flats"
Proc Int Symp on trial embankments as Malaysia Marine Clays, Kuala Lumpur
(R R Hudson, C T Toh and Chan, eds)
Malaysia Highway Authority, Vol 2, pp 53–66

Murayama, S and Ichimoto, E (1985)
"Sand compaction pile method (Compozer method)"
In: *Recent developments in ground improvement techniques*, Proc Int Symp, Bangkok
AA Balkema, Rotterdam, pp 79–84

Nakayama, J, Ichimoto, E, Kamada, H and Taguchi, S (1973)
"On stabilisation characteristics of sand compaction piles"
Soils and Foundations, Vol 13, No 3, September, pp 61–68

Wallays, M (1985)
"Deep compaction by casing driving"
In: *Recent developments in ground improvement techniques*, Proc Int Symp, Bangkok
AA Balkema, Rotterdam, pp 39–51

4.6 COMPACTION PILES

4.6.1 Definition

Compaction piles are driven piles whose primary purpose is to densify loose granular ground by their installation.

4.6.2 Principle

Driven piles compact the ground by displacement of material equal to the pile volume, and by the effects of vibration during driving. Figure 4.23 (a) gives an indication of the amount of compaction caused by a casing driven into the ground. Figure 4.23 (b) shows the effect of forming a Franki-type expanded base pile.

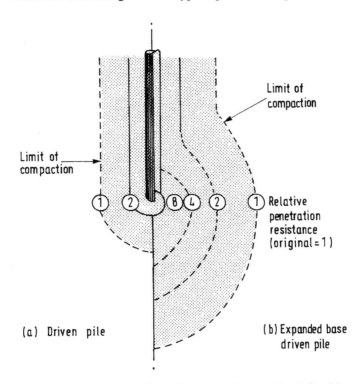

Figure 4.23 *Compaction of loose sand near piles (after Meyerhof, 1959 and 1960)*

4.6.3 Description

There are two basic forms of compaction pile. One uses a hollow steel mandrel or casing, which is driven into the ground to the depth required. The tube is closed-ended with a flap valve. Either sand or gravel backfill is then placed inside the casing as it is withdrawn slowly; the compaction of the sand or gravel column being achieved using either compressed air or a hammer working within the casing. Figure 4.24 (a) shows the process using air pressure; it is similar to the vibro-compozer system (Section 4.5.3). Figure 4.24 (b) illustrates the process using a drop hammer working within the casing – in effect constructing a Franki-type pile of aggregate rather than concrete.

The other method is by driving preformed piles into the ground on a triangular or square grid to a predetermined depth, ie the piles stay in the ground. Examples of both methods are given in Figure 4.24.

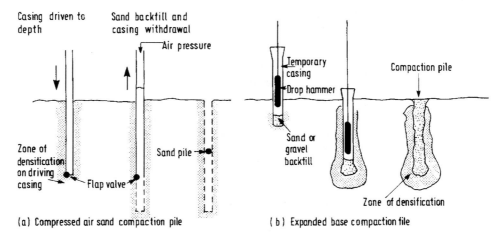

Figure 4.24 *Installation of compaction piles*

4.6.4 Applications and limitations

Mitchell (1970) advises that compaction piles can be used for soils with a fines content up to just over 20 per cent. There is a limiting depth of about 15 m. Sand compaction piles can also act as drainage channels in layered ground.

Mitchell (1970) listed early Early case histories of both sand and concrete piles being used to improve ground. Sand piles were used by Basore and Boitano (1968) to improve the seismic behaviour of 9 m of loose sand fill over 8 m of very soft mud in San Francisco. The effectiveness of vibro-compaction and compaction piles was also compared.

Moh *et al* (1981) and Moh *et al* (1982) describe two further case histories where sand compaction piles were employed to densify fine sands susceptible to liquefaction. Compaction piles were chosen because the high silt content was felt to preclude the use of vibro-compaction. Compaction piles were used in conjunction with pre-loading for one of the projects. Further case histories are reported by Loh (1982) of the expanded base technique, by Lyengar (1984) of the process using air pressure shown in Figure 4.24a, and by Jarvio and Petaja (1984) of the uses of tapered precast concrete piles for the densification of a deposit of fine sand.

Solymar and Reed (1986) also report the use of the air-pressure method on three sites where the effectiveness of compaction piling was compared with vibro-compaction, impact (ie dynamic) compaction and deep blasting. One site is described in more detail by Solymar *et al* (1984). Similar comparisons had been made earlier by Mitchell (1968 and 1970), Basore and Boitano (1968), and Schroeder and Byington (1972).

4.6.5 Design

Brons and De Kruijff (1985) suggest that compaction piles act in two ways:

- they carry a major part of the load increase, supported by the surrounding ground
- columns of ground are subject to similar vertical deformation, resulting in a stress concentration on the compaction piles.

4.6.6 Controls

Control of the treatment and workmanship is likely to involve a combination of check tests on the piling system and before and after *in-situ* tests on the treated ground, as in Section 4.5.6.

4.6.7　References

Basore, C E and Boitano, J D (1968)
"Sand densification by piles and vibro-flotation"
J Soil Mech and Found Div, Am Soc Civ Engrs, Vol 95, SM6, November, pp 1303–1323

Brons, K F and De Kruijff, H (1985)
"The performance of sand compaction piles"
Proc 11th Int Conf Soil Mech and Found Engg, San Francisco, pp 1683–1686

Jarvio, E and Petaja, J (1984)
"Improvement of the bearing capacity of underwater marine sand strata by compaction piling"
Proc 8th Eur Conf Soil Mech and Found Engg, Helsinki, 1983
AA Balkema, Rotterdam, Vol 2, pp 851–856

Loh, A K (1982)
"Soil improvements with stone columns for foundations of an oil tank"
Proc 7th Southeast Asia Geotech Conf, Hong Kong, Vol 1, pp 585–598

Lyengar, M (1984)
"Improvement of a cohesionless deposit to support a DMT process building in a seismic area"
Proc 8th Eur Conf Soil Mech and Found Engg, Helsinki, 1983
AA Balkema, Rotterdam, Vol 1, pp 55–58

Meyerhof, G G (1959)
"Compaction of sands and bearing capacity of piles"
J Soil Mech and Found Div, Am Soc Civ Engrs, Vol 85, SM6, pp 1–25

Meyerhof, G G (1960)
"The design of Franki piles with special reference to pile groups in sand"
Proc Symp on Design of Pile Foundations, 6th Int Cong Bridges and Struct Engg, Stockholm

Mitchell, J K (1968)
"Other improvement techniques (additives, grouting, thermal, electro, vibroflotation, blasting, compaction piles)"
Proc Specialty Conf on Placement and improvement of soil to support structures
MIT, Cambridge, Mass, August 1968, Am Soc Civ Engrs

Mitchell, J K (1970)
"In-place treatment of foundation soils"
J Soil Mech and Found Div, Am Soc Civ Engrs, Vol 96, SM1, January, pp 73–110

Moh, Z C, On, C D, Woo, S M and Yu, K (1981)
"Compaction sand piles for soil improvement"
Proc 10th Int Conf Soil Mech and Found Engg, Stockholm, 1981
AA Balkema, Rotterdam, Vol 3, pp 749–752

Moh, Z C, Woo, S M and Yu, K (1982)
"Pretreatment of foundation soils for oil storage tanks"
Proc 7th Southeast Asian Geotech Conf, Hong Kong, Vol 1, pp 599–614

Schroeder, W L and Byington, M (1972)
"Experiences with compaction of hydraulic fills"
Proc 10th Ann Symp on Engineering Geology and Soil Engineering, Moscow, Idaho,
pp 123–135

Solymar, Z V and Reed, D J (1986)
"Comparison between *in-situ* test results"
In: *Use of in-situ tests in geotechnical engineering,* Proc Spec Conf, Virginia,
Geotechnical Special Publication 6, Am Soc Civ Engrs, pp 1236–1248

Solymar, Z V, Hoabachie, B C, Gupta, R C and Williams, L R (1984)
"Earth foundation treatment at Jebba dam site"
J Geotech Engg Div, Am Soc Civ Engrs, Vol 110, No 10, October, pp 1415–1430

4.7 BLASTING

4.7.1 Definition

Blasting is the use of buried explosives to cause the densification of loose cohesionless ground. It has been practised in the United States for more than 40 years.

4.7.2 Principle

Detonation of the explosives in a predetermined pattern causes liquefaction, followed by the expulsion of pore water and subsequent densification of the ground (Mitchell, 1970). Gas and water escape to the surface forming sand boils, but cratering can be avoided by a suitable arrangement of the explosives.

4.7.3 Description

The usual procedure for blasting is as follows:

- jet or otherwise install a pipe to the required depth
- home the explosive charge
- withdraw the pipe
- backfill the hole
- fire the charge.

Three to five detonations are usual. Figure 4.25 illustrates ground settlement as a function of the number of charges, based on Prugh (1963), who also suggested that the first firing (marker 1) caused 50 per cent settlement, the second (marker 2) 25 per cent, the third 15 per cent, the fourth 5 per cent. Kummeneje and Eide (1961) contradict this opinion, finding that similar settlements occurred after each detonation (Figure 4.26). A typical firing pattern for pad footings is shown on Figure 4.27. To assist the densification process, a 1 m surcharge should be used in conjunction with the blasting, but the upper 1–2 m of the ground is not compacted and will require replacement and compaction in layers using a vibrating roller.

Mitchell (1970) suggested that piezometers should be used to monitor pore-water pressures during the blasting operations.

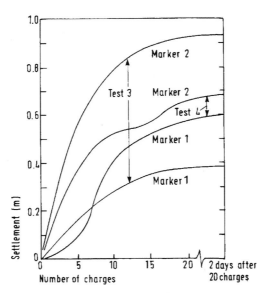

Figure 4.25 *Ground settlement as a function of the number of charges (after Prugh, 1963)*

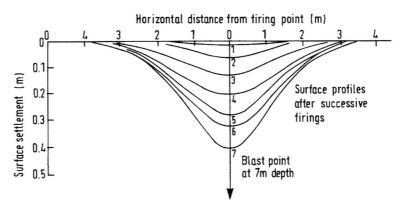

Figure 4.26 *Surface settlements after successive firings (after Kummeneje and Eide, 1961)*

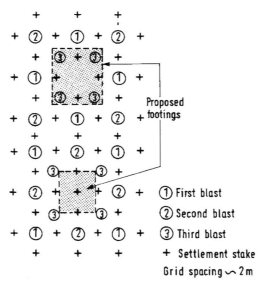

Figure 4.27 *Typical firing pattern (after Mitchell, 1970)*

4.7.4　Applications and limitations

Blasting is not usually effective in soils with contents of 20 per cent or more of silt or more than 5 per cent of clay. Increases in relative density of 15–30 per cent are claimed following blasting.

Field trials are essential, as there are no generally accepted rules. An example of this approach is by Klohn *et al* (1982), who describe a trial on sand tailings. SPT and CPT comparisons have been used to assess effectiveness of treatment. Surface settlement has to be monitored, together with pore-water pressures. It is common for there to be a substantial surface subsidence without corresponding differences in the *in-situ* test results carried out shortly afterwards. Mitchell (1982) suggests there is a healing mechanism, as a new soil structure is formed after the blasting. Blasting can destroy stratification and substantially decrease the horizontal permeability of the ground. Zones of ground originally dense can be permanently loosened or weakened.

Although not used in the UK because of the lack of suitable ground, blasting continues to be used worldwide. Case histories include Dembicki and Kisielowa (1984), where an area of about 100 000 m^2 was compacted for an ore storage yard. About 5 per cent improvement in stiffness was measured. Solymar (1984) and Solymar *et al* (1984) describe the combined use of ground improvement methods for a dam in Nigeria. Vibro-compaction was applied to 25 m depth, with blasting used from 25 m to about 40 m depth. Field trials were carried out using 0.025–0.035 kg/m of dynamite to give a measured increase in relative density of 5–15 per cent.

Barendsen and Kok (1983) describe the use of blasting in the prevention and repair of flow-slides of underwater slopes and shorelines of harbours and channels in AmsterdAm

Carpentier *et al* (1985) describe the field trials and contract work for the foundations to a new breakwater for the outer harbour at Zeebrugge, Belgium. Blasting was in competition with the Y-Probe (Section 4.5.3).

Solymar and Reed (1986) compare blasting with vibro-compaction, dynamic compaction and compaction piles for three sites. They noted that deep blasting is suited for partly to fully saturated clean loose sands and saturated silts. A relative density of 65–70 per cent was considered the maximum consistently attainable. Blasting can also be used effectively in conjunction with inundation (described in Section 5.3).

More recently, blasting was used for compaction prior to the construction of a replacement highway in the Mount St Helens area (Daniels, 1993). To bridge volcanic debris flow up to 40 m deep, zones for spread footings were compacted by blasting as were sections where the road crossed liquefiable sands. The settlement generated by the blasting in the debris flow was 1.5 m.

At the Sainte Marguerite 3 dam site in Quebec, blasting was used to densify a 16 m deep loose river sand to reduce the risk of liquefaction (Anon, 1995). Settlements of up to 1 m were achieved, and increases in relative density were measured by SPT and CPT. An important point noted by the engineers dealing with this work was that the SPT and cone resistance values continued to increase with time after the blast; thus the measurements at 12 days were considerably greater than those after two days, and those at 35 days were substantially greater again.

4.7.5   Design

Mitchell (1982) suggests the following guidelines:

- charge size: from about 1 kg to 12 kg
- depth of burial: more than a quarter of the depth to bottom of layer, $^1/_2$ to $^3/_4$ depth is common ($^2/_3$ depth is suggested by Prugh, 1963)
- charge spacing in plan: 5–15 m
- number of coverages: between one and five, with two or three usual. Each cover uses several individual charges fired successively, separated by hours to days
- total explosive use: 0.008–0.15 kg/m³, with 0.01–0.03 kg/m³ as typical
- surface settlement: 2–10 per cent of layer thickness
- small amounts of liberated gas can cause much damping of the shock waves.

Barendsen and Kok (1983) give useful design guidance in situations where it is important not to form a crater or to generate surface waves that could damage nearby structures. In these situations, the overall average settlements in sands will be typically about 2 per cent of the effective depth treated, depending on its initial relative density.

4.7.6   Controls

In addition to the *in-situ* tests such as SPT or CPT soundings before and after treatment, the critical controls will be those on charge weights, depths and timing, not only in relation to attempted densification, but also to any environmental restrictions, such as vibration or noise. Note the different requirements of whether crater formation can be permitted, and the observation that testing post treatment should not be too early, as the penetration test results for sands in particular appear to increase for a considerable time.

4.7.7   References

Anon (1995)
"Explosive issue"
Ground Engineering, July/August, pp 24–25

Barendsen, D A and Kok, L (1983)
"Prevention and repair of flow-slides by explosion densification"
Proc 8th European Conf Soil Mech and Found Engg, Helsinki
AA Balkema, Rotterdam, Paper 3.4, pp 205–208

Carpentier R, DeWolf, P, Van Damme, L, De Rouck, J and Bernard, A (1985)
"Compaction by blasting in offshore harbour construction"
Proc 11th Int Conf Soil Mech and Found Engg, San Francisco
AA Balkema, Rotterdam, Vol 3, pp 1687–1692

Daniels, S (1993)
"Building across volcanic debris"
Engineering News Record, 2 August, p 25

Dembicke, E and Kisielowa, N (1984)
"Technology of soil compaction by means of explosion"
Proc 8th Eur Conf Soil Mech and Found Engg, Helsinki, 1984
AA Balkema, Rotterdam, Vol 1, pp 229–230

Klohn, E J, Garga, V K and Shukin, W (1982)
"Densification of sand tailings by blasting"
Proc 10th Int Conf Soil Mech and Found Engg, Stockholm, 1981
AA Balkema, Rotterdam, Vol 3, pp 725–730

Kummeneje, O and Eide, O (1961)
"Investigation of loose sand deposits by blasting"
Proc 5th Int Conf on Soil Mech and Found Engg
Dunod, Paris, Vol 1

Mitchell, J K (1970)
"In-place treatment of foundation soils"
J Soil Mech and Found Div, Am Soc Civ Engrs, Vol 96, SM1, January, pp 73–110

Mitchell, J K (1982)
"Soil improvement – state of the art"
Proc 10th Int Conf Soil Mech and Found Engg, Stockholm, 1981
AA Balkema, Rotterdam, Vol 4, pp 509–565

Prugh, B J (1963)
"Densification of soils by explosive vibrations"
J Constn Div, Am Soc Civ Engrs, Co1, March, pp 79–100

Solymar, Z V (1984)
"Compaction of alluvial sands by blasting"
Can Geot J, Vol 21, pp 305–321

Solymar, Z V, Iloabachie, B C, Gupta, R C and Williams, R L (1984)
"Earth foundation treatment at Jebba dam site"
J Geotech Engg Div, Am Soc Civ Engrs, Vol 110, No 10, October, pp 1415–1430

Solymar, Z V and Reed, D J (1986)
"Comparison between *in-situ* test results"
In: *Use of in-situ tests in geotechnical engineering*, Proc Spec Conf, Virginia
Geotech Special Publication 6, Am Soc Civ Engrs, pp 1236–1248

5 Improvement by adding load or increasing effective stress

Increasing the load on the ground causes it to compress. How much compression and how long it takes to happen depends on the arrangement of the ground particles, on the degree of saturation, and on how freely the soil can drain. For loose and particularly unsaturated fills, adding load induces rapid settlement; soft, saturated clays, on the other hand, take months or years to consolidate under an added load while pore pressures dissipate and the effective stress in the soil increases.

The improvement techniques described here fall into two, not necessarily exclusive, types:

- where the improvement largely comes about by the increase in total stress
- where the improvement depends upon the increase in effective stress and the technique encourages or accelerates that.

The techniques are presented in the following order:

- pre-compression (Section 5.1)
- vertical drains (Section 5.2)
- inundation (Section 5.3)
- vacuum pre-loading (Section 5.4)
- dewatering (Section 5.5)
- electro-osmosis (Section 5.5.4)
- pressure berms (Section 5.6).

Some of these methods are rarely used – if at all – in the UK. Others, eg vertical drains, appear to be more widespread because they are given extensive coverage in the technical literature and in the technical press.

Dewatering is probably the most frequently used method of ground improvement in temporary works for the control of groundwater into excavations. The general principles set out in Chapters 1 to 3 of the approach to its selection, design and operation all apply. It is, however, used only rarely for "permanent" ground improvement.

5.1 PRE-COMPRESSION

5.1.1 Definitions

Pre-compression is the deliberate act of compressing the ground under an applied pressure before placing or completing the structural load. Pre-compression takes two forms (Aldrich, 1964):

- *pre-compression* is achieved by pre-loading, which requires placement and removal of earth, water, or some other dead load, before construction, similar in weight to the structural load (or final load)
- *surcharging* is where the stress intensity from the pre-loading is greater than the intensity from the final loading.

Surcharge is the excess load intensity above final load, and the *surcharge ratio* is the ratio of surcharge to final load.

5.1.2   Principles

Pre-compression is used to induce settlements that have three recognisable phases: immediate, primary consolidation and secondary consolidation (or creep). These are illustrated in Figure 5.1, the proportions being typical of a soft clay. In a loose fill with large voids, the potential for immediate settlement can be a very high proportion. The effects of pre-compression on each phase of the settlement of the finished structure load are beneficial and are summarised in Table 5.1.

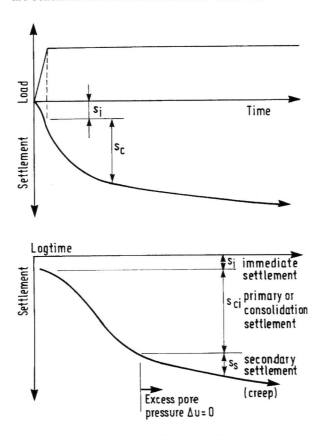

Figure 5.1   *Types of settlement (after Jamiolkowski et al, 1984)*

Table 5.1   *Effect of pre-compression on phases of settlement*

Settlement phase	Effect of pre-compression
1. Immediate	Minimises amount
2. Primary consolidation	Reduces amount
3. Secondary consolidation or creep	Reduces rate

By pre-compression with a surcharge the primary settlements may be compensated (in the sense that what would be the settlement under the final load is reached earlier): this is shown on Figure 5.2. To control rebound on removal of surcharge, the final load, p_f, should not be less than one-third of the surcharge load, ie $p_s/3$. The time for removing the surcharge from a soft clay is often taken as when the average degree of consolidation over the clay layer, U, is at least $p_s/(p_s + p_f)$ and typically 75 per cent or so.

Secondary settlements can be partly compensated by maintaining the surcharge for some time past its 100 per cent primary consolidation before removal (see Bjerrum, 1972 and Johnson, 1970).

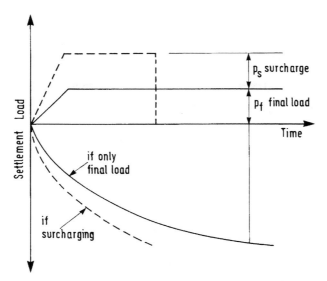

Figure 5.2   *Compensating primary settlement by surcharging (after Jamiolkowski et al, 1984)*

Step-by-step (or stage) construction is used where the ground is so weak that it has to gain sufficient strength before the next lift of pre-load or surcharge can be added (see Figure 5.3).

5.1.3   Description

Pre-loading and surcharging usually involve placing an earth fill over quite a large area or over several areas within one site, the technique being more economical on moderate-to-large sites (Broms, 1979). For full economy the fill should be reused. The loaded area has to be larger than that of the proposed development. Materials to form the load can be earth- or rock-fill, water (Tozzoli and York, 1973) or any other easily transportable and available bulk fill or weighty material.

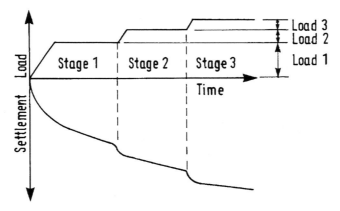

Figure 5.3   *Stage construction*

5.1.4   Applications

The pre-compression method is applied to many types of ground conditions, but usually to soft clays, organic clays, peats, loose silts, fine sands and fills, including refuse fills. Aldrich (1964) describes how the use of pre-compression developed at the end of the 1940s for light structures, oil storage tanks and embankments.

Charles *et al* (1986) describe nine case histories of pre-loading of uncompacted fills (including opencast backfill) on derelict sites. They suggest that pre-loading is particularly effective because the fills are unsaturated, highly permeable and so compress rapidly, so that the majority of settlement occurs during the placing of the surcharge itself, as in Figure 5.4. This contrasts with the slower consolidation of saturated silts and clays.

Charles *et al* (1986) note that all forms of settlement discussed in Appendix A have to be considered, as well as the possibility of gas generation. They recommend that $p_s > 1.2\ p_f$.

In a more recent case history, Jones *et al* (1987) describe how a 2 m surcharging was applied to deep deposits of soft clays and peats to allow the construction of up to 4 km of roads, drainage and services. In their appraisal they noted that the cost of surcharging would be less than half that of pre-loading with vertical drains, but the development programme has to be extended to undertake the work in phases. The fill material was used three times in surcharging and later in a permanent landfill.

Burford (1991) describes the monitoring of a 10 ha site of restored deep opencast backfill. The fill is about 24 m deep. The surcharge imposed a bearing pressure of 150 kN/m². Some 300 mm of settlement took place within the upper 12 m. Houses were built after removal of the surcharge.

5.1.5 Limitations

The definitive work on pre-compression is probably that by Johnson (1970). He points out that the promoter ought to be made aware of the risks of using this technique. According to York (1968), these include:

- insufficient time available to achieve the required compression or pore-pressure dissipation

- potential shear failure or excessive lateral deformation of the pre-compression fill, often because of the transmission of high pore pressure beyond its toe along thin permeable layers

- inadequate factor of safety against shear failure when the structure is complete. It is essential to check stability at this stage

- post-construction settlements may be greater than expected.

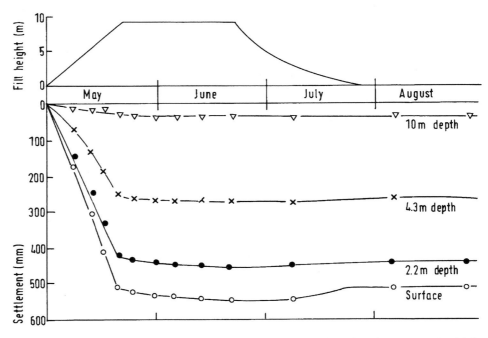

Figure 5.4   *Settlements of a surcharged opencast backfill (after Charles* et al, *1986)*

For these reasons, the ground investigation has to be very thorough and a field trial loading is essential.

The main constraint for saturated clay soils is the available time. When loose, open-textured fills are pre-compressed, however, a great deal of the settlement takes place very quickly, and the available time should not be a constraint.

The pre-compression fill should either be free draining or, if on impermeable ground, a drainage layer should be laid first. Note that loose, clean sands are not densified very much by pre-compression.

5.1.6   Design

While the concept of pre-compression is simple, design is often intuitive (Hansbo, 1984) because the consolidation solutions can be complex, often three-dimensional, and because of uncertainties of the ground conditions.

A critical determinant is the pre-consolidation pressure, p_c, of the soil. If the final load, p_f, is less than this, there is no need for pre-loading. On the other hand, pre-compression will only be effective if $(p_f + p_s)$ is very much greater than p_c.

A good understanding of the ground conditions and the consolidation and permeability characteristics of the fine-grained compressible soils is needed. Because time is the principal consideration, deep drains (Section 5.2) are often used in conjunction with pre-compression to accelerate the rate of settlement.

When pre-compression is used for building developments on natural ground, the surcharge ratio (p_s/p_f) is typically between 0.5 and 1.0 and the loading period from one to six months. On fills, Charles et al (1986) recommend a ratio of more than 1.2. For embankments, surcharge ratios are 0.3 or less and the load applied for several months or even one or two years. The heights of pre-compression fill are rarely as much as 10 m. On the other hand, Jones et al (1995), following up the earlier paper Jones et al (1987), record how they applied considerably higher ratios of 2 to >6 (their term "load ratio" is the same as "surcharge ratio") to construct a road over 12–25 m of very compressible clays, silts and peat in coastal marshland in South Wales. Here the surcharge had a 25 m crest width, 1 in 4 side slopes and a maximum height of 4.5 m. They explain that the soil model used was based on conventional consolidation parameters and the performance over the long term was as predicted.

5.1.7   Controls

The principal criterion is to consolidate the ground so that when it bears the project loading there is little further settlement. Thus the main indicator of the effectiveness of the treatment is the amount of settlement at the surface and, possibly also, at depths throughout the treated zone. This is not a sufficient control, however, as the settlement may be rapid or prolonged or there may be, as with soft soils, potential foundation instability. Control over the rate of filling, over pore pressure responses in the foundation soils, and perhaps over lateral movements in the foundation may all be necessary.

5.1.8   References

Aldrich, H P (1964)
"Precompression for support on shallow foundations"
In: *Design of foundations for control of settlement*
Am Soc Civ Engrs, pp 471–486

Bjerrum, L (1972)
"Embankments on soft ground"
In: *Performance of earth and earth-supported structures*
Proc Conf West Lafayette, Indiana, Am Soc Civ Engrs, Vol II, pp 1–54

Burford, D (1991)
"Surcharging a deep opencast backfill for housing development"
Ground Engineering, Vol 24, No 7, September, pp 36–39

Broms, B B (1979)
"Problems and solutions to construction in soft clays"
Proc 6th Reg Conf Soil Mech and Found Engg, Singapore, Vol 2, pp 27–36

Charles, J A, Burford, D and Watts, K S (1986)
"Improving the load carrying characteristics of uncompacted fills by preloading"
Municipal Engineer, Vol 3, No 1, pp 1–19

Hansbo, S (1984)
"Techno-economic trend of subsoil improvement methods in foundation engineering"
Special lecture, *Proc 8th Eur Conf Soil Mech and Found Engg*, Helsinki, 1983
AA Balkema, Rotterdam, Vol 3, pp 1333–1343

Jamiolkowski, M, Lancellotta, R and Wolski, W (1984)
"Precompression and speeding up consolidation"
Proc 8th Eur Conf Soil Mech and Found Engg, Helsinki, 1983
AA Balkema, Rotterdam, Vol 3, pp 1201–1226

Johnson, S J (1970)
"Precompression for improving foundation soils"
J Soil Mech and Founds Div, Am Soc Civ Engrs, Vol 96, SM1, 111–170

Jones, D B, Beasley, D H and Pollock, D J (1987)
"Ground treatment by surcharging on deposits of soft clay and peat: a case history"
In: *Building on marginal and derelict land*
Thomas Telford, London, pp 679–695

Jones, D B, Maddison, J D and Beasley, D H (1995)
"Long-term performance of clay and peat treated by surcharge"
Geotechnical Engineering, Proc Instn Civ Engrs, Vol 113, Issue 1, pp 31–37

Tozzoli, A J and York, D L (1973)
"Water used to pre-load unstable subsoils"
Civil Engineering, Am Soc Civ Engrs, August, pp 56–59

York, D L (1968)
"Densification after placement (drains)"
In: *Placement and improvement of soil to support structures*
Am Soc Civ Engrs, pp 5–7

5.2 VERTICAL DRAINS

5.2.1 Definition

Vertical drains are long, thin drainage elements installed vertically through relatively impermeable, soft clay soils to accelerate the speed at which they consolidate when loaded. Vertical drains are often used in conjunction with pre-compression (Section 5.1). Other names and variations of this technique are: deep drains, sand drains, sandwicks, band drains, wick drains or prefabricated band-shaped drains.

5.2.2 Principle

The rate of consolidation of a soil, whether in terms of dissipation of excess pore pressure or settlement, depends on the square of the length of the shortest drainage path. Vertical drains are installed so as to reduce the natural drainage path distances, usually exploiting the naturally higher horizontal permeability of clay deposits, particularly those with sand or silt layers and lenses (Figure 5.5). There is thus a combination of radial and vertical pore-water flow to the drains, each of which is considered as being within a cylinder of soil (Figure 5.6). Vertical and radial consolidation can be calculated separately from theoretical considerations, but the overall performance of a vertical-drain system requires assessment of the effects of drain installation (remoulding and smear of the surrounding clay), of the consolidation and flow characteristics of the soil in relation to its depositional structure and macrofabric, and to the flow characteristics of the drain and how these can change.

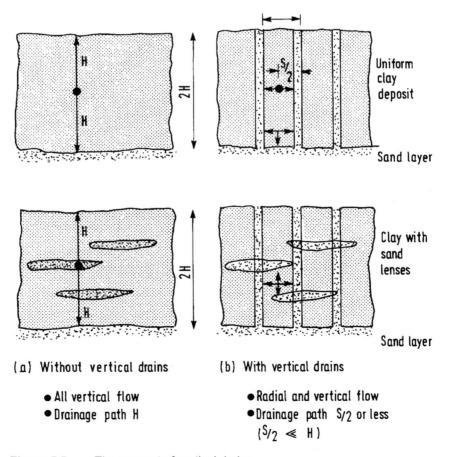

(a) Without vertical drains

- All vertical flow
- Drainage path H

(b) With vertical drains

- Radial and vertical flow
- Drainage path $S/2$ or less
($S/2 \ll H$)

Figure 5.5   *The concept of vertical drains*

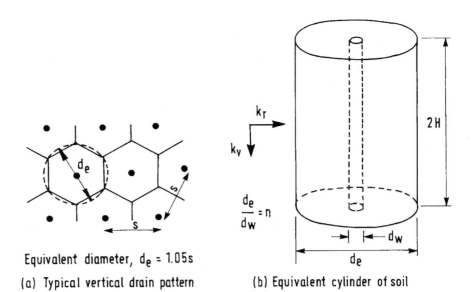

Equivalent diameter, $d_e = 1.05s$

(a) Typical vertical drain pattern

(b) Equivalent cylinder of soil

Figure 5.6 *Vertical drain geometry*

5.2.3 Description

The attributes required of vertical drains were listed by Hughes and Chalmers (1972) as:

- high permeability to permit rapid dissipation of pore-water pressures

- sufficient flexibility to accept large vertical and lateral ground movements. A single drain should not act as a pile, inhibiting consolidation. It should have about the same stiffness as the surrounding soil mass

- continuity over its full length and a good hydraulic connection with the drainage blanket at ground level which acts as a hydraulic sink

- an installation method that does not cause so much disturbance as to make the surrounding soil too impermeable for the drain to be effective

- ability to function over the required period, which may be for a few months to up to two years

- adequate drain characteristics in changing conditions of stress, usually increasing stress

- filters that do not become clogged by the surrounding fine-grained soils.

In the 1930s deep drains were first used in the USA by filling 20 inch-diameter borings with sand. Independently, cardboard wick drains were developed at the same time in Sweden (Kjellman, 1948).

In the late 1960s the sandwick drain came into use as a smaller form of sand drain. Hughes and Chalmers (1972) describe the sandwick as a fabric stocking filled pneumatically with sand to a diameter of 60–70 mm.

Building on Kjellman's proposals, many geotechnical companies in the early 1970s developed band or wick drains. These are prefabricated using modern polymer materials (eg polyethylene, polypropylene) for the core, and woven or non-woven fabrics, fibre or paper for the filter.

On large sites with closely spaced and often deep drains, the quantities in terms of total length of drain and number can be high. Rapid, efficient installation is a critical cost factor. McGown and Hughes (1982) classified the installation methods under three categories: displacement methods, drilling methods and washing (or jetting) methods.

Prefabricated drains (including sandwicks) all use a closed-end mandrel for installation, the degree of soil displacement depending on their geometry. Such drains can be installed by displacement, drilling or jetting methods. Figure 5.7 compares the installation methods and their effects on displacing and smearing the soil around the drain.

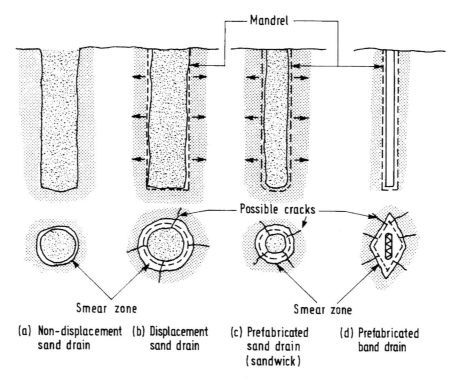

Figure 5.7   *Types of drain installation and effects*

Table 5.2 compares main types of vertical drains, their installation methods, diameter (d_w) and usual spacing, *s*. Maximum installation depths are usually less than 30 m although band drains can be installed to depths approaching 60 m. Some of the many types of prefabricated band-shaped drains are listed in Table 5.3 showing their basic dimensions and the materials from which the filter and core are formed.

Sand drains, sandwicks and band drains are described in more detail below.

Sand drains

Displacement sand drains (Figure 5.7b) are cheap and simple to install, but cause the most disturbance and borehole smear (Casagrande and Poulos, 1969). As well as reducing horizontal permeability locally, the disturbance increases pore-water pressures and decreases shear strength in the surrounding soil, the effect sometimes being self-defeating. Sand drain diameters are typically in the range 200–600 mm, so driving large groups causes substantial ground heave and lateral displacement (Akagi, 1981). Disturbance tends to increase with depth. Sand drains can also stiffen the ground, inhibiting settlement (see the vibro-compozer system in Section 4.5).

Displacement sand drains should not be installed in sensitive clays (ie sensitivity more than about 5) or in soils with a highly developed macrofabric (Jamiolkowski *et al*, 1984). Non-displacement installation methods are more effective in limiting disturbance (remoulding), but there will be a smear zone around the borehole walls. Washing methods may give irregularly shaped drains of unknown size. The sand should be saturated when filling the drain holes.

Table 5.2   *Types of vertical drains*

Drain type	Installation method	Typical ranges of:	
		drain diameter d_w (m)	drain spacing s (m)
Sand drains			
formed in place (1) (Figure 5.7b)	Driven closed-end mandrel	0.2–0.6	1.5–5
formed in place (2) (Figure 5.7a/b)	Continuous flight, hollow-stem auger	0.3–0.5	2–5
formed in place (3) (Figure 5.7a)	Jetted	0.2–0.3	2–5
prefabricated (sandwick) Figure 5.7c)	Driven closed-end mandrel	0.06–0.15	1.2–4
Band drains			
(Figure 5.7d)	Driven closed-end mandrel	0.05–0.1 (equivalent diameter)	1.2 to 3.5

Sandwicks

Sandwicks can be considered as prefabricated sand drains. Made of woven jute or polypropylene or of melt-bonded fabric stockings and filled with clean, sharp concreting sand, they are normally 60–70 mm in diameter. In the UK, woven polypropylene is usual. They are usually installed by driving a mandrel of about the same size as the drain (Figure 5.7b), although they can be placed in drilled or jetted holes. The sand-filled stockings can accommodate substantial vertical and lateral deformation without impairment of the drain's efficiency.

Comparisons of the performance of sandwicks with other types of vertical drain have been published by Cole and Garrett (1981), Nicholson and Jardine (1982) and Hansbo *et al* (1982).

Band (or wick) drains

There are now more than 50 drains of this type on the market. Table 5.3 lists the dimensions and materials of some of these. Band drains are installed by displacement methods. This means connecting them to a suitably shaped mandrel, which is then driven into the ground either dynamically or statically. Hansbo (1979) reports that the dynamic method, a vibrating hammer, develops far higher excess pore-water pressures than the static methods. Figure 5.8 shows a typical arrangement using a vibrator to drive the mandrel. Cross-sections of two mandrels used for a plastic and cardboard drain respectively are shown in Figure 5.9.

Table 5.3   *Dimensions and materials of some prefabricated drains (after Holtz et al, 1991)*

Drain type	Dimensions		Materials	
	Width a (mm)	**Thickness b (mm)**	**Filter**	**Core**
Kjellman	100	3.0	Cardboard	Cardboard
Alidrain	100	6.1	Geotextile	Polyethylene
Amerdrain	92	10	Geotextile	Polypropylene
Bando	96	2.0	Paper	PVC
Castleboard	93	3.2	—	Polyolefin
Colbond	300	4	Geotextile	Polyester
Desol	95	2	—	Polyolefin
Geodrain	98	4	Paper	Polyethylene
Geodrain	96	3.5	Geotextile	Polyethylene
Hitek	100	6	Geotextile	Polyethylene
Mebradrain	95	3.2	Paper	Polypropylene or polyethylene
Mebradrain	95	3.4	Geotextile	Polypropylene or polyethylene
OV Drain	103	—	Geotextile	Polyester
Solpac Charbonneau	105	—	Geotextile	Polyester
Tafnel	102	6.0	—	Polypropylene

Note: for band-shaped drains the equivalent drain diameter, $d_w = 2 \, (a + b)/\pi$

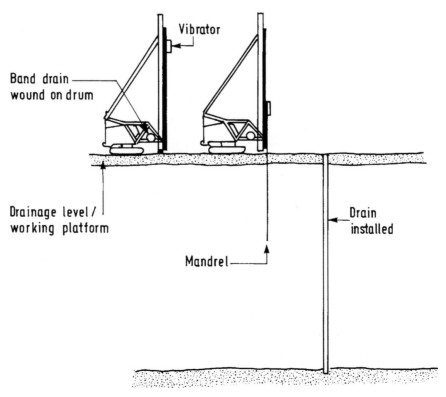

Figure 5.8   *Installation of band drains*

5.2.4   Applications

Vertical drains greatly accelerate the rate of primary or consolidation settlement of impermeable soft clays. They are, therefore, most often used in combination with pre-compression loading (Section 5.1), sometimes for large areas of reclamation or for embankments on soft clays. In addition, drains can be used to increase the stability at toes of embankments and to act as pressure relief wells for any artesian water pressures.

If the use of deep drains is being contemplated, a very thorough ground investigation is needed. Essential requirements are to investigate the macrofabric and drainage paths of the soil deposits, preferably by continuous, high-quality sampling and piezocone soundings. A combination of laboratory and field testing is needed to assess the horizontal and vertical permeabilities and compressibility. For large projects, particularly where the clay has a macrofabric, a trial embankment is the best way to assess the consolidation characteristics of the ground; it can also be used to assess different drain types or spacings. Holtz *et al* (1991) give thorough guidance on the importance of different design parameters for vertical drain systems and how to estimate them.

The need to understand soil macrofabric was emphasised by Rowe (1968) and Johnson (1970), with respect to sand drains, and McGown *et al* (1980) propose methods for describing the different sorts of macrofabric features of soils.

The preconsolidation pressure also has to be determined to understand the stress history of the ground. There is no point in using vertical drains if the permanent loading is not to be greater than the preconsolidation pressure (Poulos, 1968; Jamiolkowski *et al*, 1984).

Primary, or consolidation settlement, and secondary compression are defined on Figure 5.1. Where primary settlement is large compared with secondary compression deep drains can be useful. For soils such as peats, highly stratified or very organic clays, secondary compression is likely to be greater than primary settlement and vertical drains will not be of benefit (Broms, 1979; Akagi, 1981).

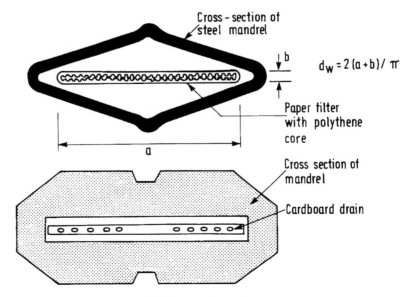

Figure 5.9   *Two band drains and mandrels*

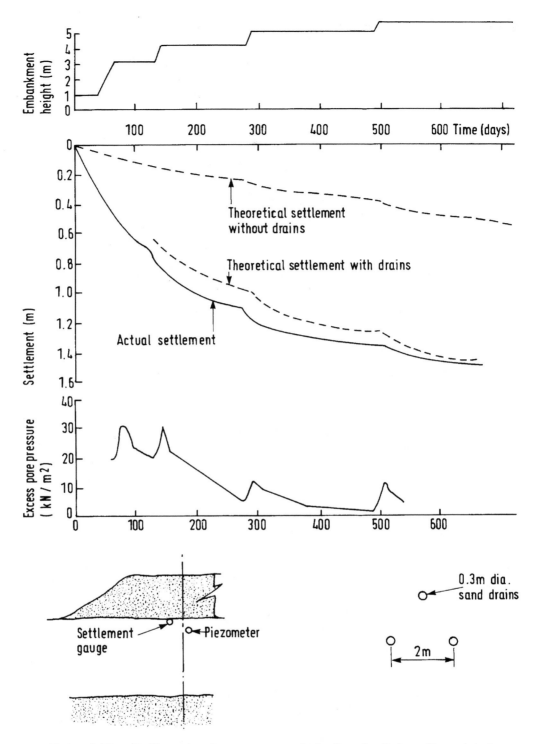

Figure 5.10 *Settlement and pore pressure dissipation of soft clay with sand drains below an embankment (after Pilot, 1978)*

A typical sand drain case history is shown on Figure 5.10, in which stage construction of about 5 m of fill was placed on about 8 m of soft clay.

Hansbo (1979) suggests that drains can be used effectively in conjunction with a vacuum, as an alternative to using fill. This is discussed in Section 5.4. Wolski *et al* (1979) also describe a site where drains were used very effectively in conjunction with dewatering an underlying permeable layer.

5.2.5   Limitations

The technique of vertical drains is appropriate to soft clays when the applied load is above the preconsolidation pressure and secondary compression will not be the main component of settlement. It can be difficult to predict the rate of consolidation reliably. Often settlements occur much faster than predicted; sometimes it turns out that even without vertical drains the consolidation of layered clays would be sufficiently rapid. The difficulty is making an accurate estimate of the consolidation and permeability characteristics of the soils and of the effects of installation on drain performance.

Fill may have to be placed in lifts (Figures 5.3 and 5.10) in order to gain sufficient foundation strength so that there is no risk of instability. Usually a granular blanket has to be placed over the site to form a working platform for the rigs. All of the drain types suffer from difficulties of installation or deterioration of performance in the ground. Jetted sand drains, although non-displacement and with little smear, are difficult to install through coarse granular layers or stiff clay fills. Drilling methods, while giving more accurate holes and usually with small displacement when the sand is placed, tend to create a wide smear zone of low permeability.

In deep holes, sand can arch across the tube and leave gaps beneath. There is a risk of soil squeezing into a gap, so making the drain discontinuous.

While wick and band drains are continuous, soil fines can clog the drain filter, whether this is a fabric stocking or the paper, cardboard or geotextile of a band drain, thereby reducing the effectiveness of the drain. If the flow channels of the drain become clogged or silt up, the drain's discharge capacity is reduced.

Band drains are simple, quick and cheap to install, but Jamiolkowski *et al* (1984) and Holtz *et al* (1991) warn that there are still unknowns regarding their mechanical, hydraulic and durability properties. There is no international standard yet available. Prefabricated drains can deteriorate under bacteriological or chemical attack, above and below the water table. The drain can be stopped from functioning within one year in organic ground. These comments apply to both paper and geotextile filters. Jamiolkowski *et al* (1984) suggest the following lives for drains:

- non-impregnated paper – 12 to 16 months
- impregnated paper – 24 to 30 months
- non-woven filter sleeves – no data.

The ground is disturbed with the installation of the drain, but the magnitude of the effect is unknown. It is largely controlled by the dimensions of the mandrel. In extreme circumstances the ground could relax around the filter, forcing it into the core channels and preventing water flow. Hansbo *et al* (1982) consider that this effect will control discharge capacity in service. Tests to determine discharge capacity are described by Holtz *et al* (1991). Band drains also need to be protected from the wind and direct sunlight (ultra violet light) or they will degrade rapidly (McGown and Hughes, 1982; Nicholls *et al*, 1984).

As the clay compresses, the drains distort, folding and perhaps kinking, so reducing the discharge capacity. The effect on discharge capacity has been simulated by crimping the band drain to various shapes and by in-soil tests after 20 per cent compression. Koda *et al* (1989) found multiple kinking of drains exposed by test pits some three to four years after installation. Here an embankment on 11 m of peats and very soft calcareous clays settled about 2 m. Discharge capacities of the drains, although much reduced, were still acceptable at about 100 m^3/year.

5.2.6 Design

For the design of sandwicks, guidance is given by Davies and Humpheson (1982), and by Nicholson and Jardine (1982) in comparative trials with band drains. Hansbo *et al* (1982), Jamiolkowski *et al* (1984) and Holtz *et al* (1991) describe in detail the design of deep drains in general and band drains in particular.

5.2.7 Controls

Installation effects can jeopardise vertical drain efficiency, eg smearing, so care is needed to check that the installed drains work as expected. Other construction controls are drain spacing and depth and checking their connection to surface drains. At the loading stage, it will usually be necessary to monitor pore pressures and settlement and, perhaps, lateral displacements of a foundation soil beneath an embankment. In some circumstances the critical criterion is a gain in undrained shear strength of the foundation, and strength profiling with a penetration vane or CPT soundings may be warranted.

5.2.8 References

Akagi, T (1981)
"Effects of mandrel-driven sand drains on soft clay"
Proc 10th Int Conf Soil Mech and Found Engg, Stockholm, 1981
AA Balkema, Rotterdam, Vol 3, pp 581–584

Broms, B B (1979)
"Problems and solutions to construction in soft clays"
Proc 6th Reg Conf Soil Mech and Found Engg, Singapore, Vol 2, pp 27–36

Casagrande, L and Poulos, S (1969)
"On the effectiveness of sand drains"
Can Geotech J, Vol 6, No 3, pp 287–326

Cole, K W and Garrett, C (1981)
"Two road embankments on soft alluvium"
Proc 10th Eur Conf Soil Mech and Found Engg, Stockholm, 1981
AA Balkema, Rotterdam, Vol 1, pp 87–94

Davies, J A and Humpheson, C (1982)
"A comparison between the performance of the two types of vertical drain beneath a trial embankment in Belfast"
In: *Vertical drains*
Thomas Telford, London, pp 19–31

Hansbo, S (1979)
"Consolidation of clay by band-shaped prefabricated drains"
Ground Engineering, Vol 12, No 5, pp 16–25

Hansbo, S, Jamiolkowski, M and Kok, L (1982)
"Consolidation by vertical drains"
In: *Vertical drains*
Thomas Telford, London, pp 45–66

Holtz, R D, Jamiolkowski, M B, Lancellotta, R and Pedroni, R (1991)
Prefabricated vertical drains: design and performance
Book 11, CIRIA, Westminster, and Butterworth-Heinemann, Oxford

Hughes, F H. and Chalmers, A. (1972)
"Small-diameter sand drains"
Civ Engg Publ Wks Rev, Vol 67, 788, pp 3–6

Jamiolkowski, M, Lancellotta, R and Wolski, W (1984)
"Precompression and speeding up consolidation"
Proc 8th Eur Conf Soil Mech and Found Engg, Helsinki, 1983
AA Balkema, Rotterdam, pp 1201–1226

Johnson, S J (1970)
"Foundation precompression with vertical sand drains"
J. Soil Mech and Found Div, Am Soc Civ Engrs, Vol 96, SM1, pp 145–157

Kjellman, W (1948)
"Accelerating consolidation of fine grained soils by means of cardboard wicks"
Proc 2nd Int Conf Soil Mech and Found Engg, Rotterdam, Vol II, pp 302–305

Koda, E., Szymanski, A. and Wolski, W. (1989)
"Behaviour of geodrains in organic subsoil"
Proc 12th Int Conf Soil Mech and Found Engg, Rio de Janeiro, Vol 2, pp 1377–1380

McGown, A and Hughes, F H (1982)
"Practical aspects of the design and installation of deep vertical drains"
In: *Vertical drains*
Thomas Telford, London, pp 3–17

McGown, A, Marsland, A, Radwan, A M and Gabr, A W A (1980)
"Recording and interpreting soil microfabric data"
Géotechnique, Vol 30, No 4, pp 417–447

Nicholls, R A, Barry, A J and Shoji, H (1984)
"Deep vertical drain installation"
Ground Engineering, Vol 17, No 4, May, pp 31–35

Nicholson, D P and Jardine, R J (1982)
"Performance of vertical drains at Queenborough Bypass"
In: *Vertical drains*
Thomas Telford, London, pp 67–90

Pilot, G (1978)
"State of the art"
Bulletin de Liaison des Laboratoires des Ponts et Chaussées, Soil Mechanics, Special
Issue VI, April, pp 140–178

Poulos, S J (1968)
"Densification after placement (drains)"
In: *Placement and improvement of soil to support structures*
Am Soc Civ Engrs, pp 43–52

Rowe, P W (1968)
"The influence of geological features of clay deposits on the design and performance of sand drains"
Proc Instn Civ Engrs, Paper 7058-S, pp 1–72

Wolski, W, Furstenberg, A and Lechowicz, Z (1979)
"Consolidation parameters of a soft clay layer improved by means of vertical drains"
In: *Design parameters in geotechnical engineering*
Proc 7th Eur Conf Soil Mech and Found Engg, Brighton, 1979, British Geotech Soc, London, Vol 3, pp 311–314

5.3 INUNDATION

5.3.1 Definition

Inundation, as a form of ground improvement, is the application of water to loose unsaturated ground causing collapse settlement and making the ground more dense.

5.3.2 Principle

Many natural soils worldwide have a loose honeycomb structure with various types of bonding that hold the grains in place, as shown in Figure 5.11. Inundation causes collapse of the grain structure by breaking or destroying these bonds. Figure 5.12 illustrates the principle of collapse settlement: it occurs on wetting at no change of effective stress. Collapsible natural ground includes windblown deposits of sands and silts, loess, alluvial flood plains, and colluvial or residual tropical soils, of the kind described by Mitchell (1982). Such natural ground is largely absent in the UK.

However, Clayton (1980) and Stroud and Mitchell (1990) describe how collapse settlement can occur in chalk fills. There are many loose backfills of quarries and old opencast sites that are susceptible to collapse settlement on inundation, some of which effects have been studied by BRE (Charles *et al*, 1979; Charles *et al*, 1985).

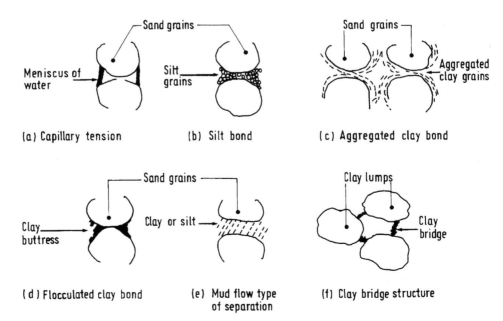

Figure 5.11   *Collapsible soil mechanisms (after Mitchell, 1976)*

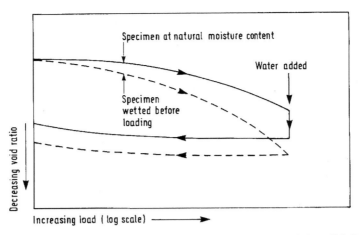

Figure 5.12   *Effect of loading and wetting a collapsible soil (after Jennings and Knight, 1957)*

5.3.3   Description

Water can be added by ponding or through trenches. Charles *et al* (1979) describe a site at Corby involving unsaturated cohesive fill. Five trenches, 1 m deep, were dug at 10 m centres across a 50 m square area, and kept filled with water for four months. Figure 5.13 shows the settlement that occurred with time and at different depths. Charles *et al* (1979) point out that for inundation to be effective the ground should be dry of its optimum moisture content. They noted that deeper, more closely spaced drains would have been more effective than shallow trenches.

5.3.4   Applications and limitations

The case history presented by Charles *et al* (1979) includes a comparison with dynamic compaction and pre-loading. It was found that inundation was the least successful (see Table 4.3). Deep deposits of loess in Russia and Bulgaria have been improved by inundation in conjunction with blasting by using deep drain wells which can then be used to place explosive charges (Litvinov, 1973; Donchev, 1980; Minkov *et al* 1981).

Charles *et al* (1985) give data for opencast sites backfilled with colliery waste fragments of mudstone and sandstone. Dewatering had been discontinued; eg at Horsley the ground water rose 34 m in three years. Vertical compressions of 1–2 per cent were caused, mainly in the upper 10 m of unsaturated backfill. Elsewhere, Charles *et al* (1979) and Smyth-Osbourne and Mizon (1985) have reported vertical compressions of 4–6 per cent.

Inundation can occur in several ways:

- by the recovery of groundwater levels after deep pumping ceases
- by leakage from water mains or by rainwater entering fills
- by deliberately recharging dewatered excavations to control settlement of surrounding ground.

The first two being uncontrolled can cause irregular, unexpected collapse settlements of fills. The third is a deliberate technique of recharge to offset the effect of dewatering of excavations. Substantial settlements beyond the excavation arise because of the increase in effective stresses. Davies and Henkel (1980) show the effect of recharge wells in controlling settlement for Chater Station in Hong Kong. The same principle was used by Humpheson *et al* (1986) for the basement to the Hong Kong and Shanghai Bank. The principle in this case is not to cause collapse settlement but to reduce the change in effective stress by keeping the pore pressure high.

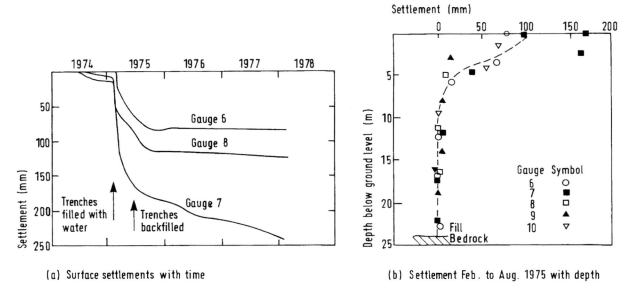

(a) Surface settlements with time

(b) Settlement Feb. to Aug. 1975 with depth

Figure 5.13 *Settlements of inundated cohesive fill (after Charles et al, 1979)*

5.3.5 Design

Field trials are essential, and in natural soils great care must be taken to carry out laboratory tests correctly. Where there is the possibility of collapse settlement occurring unpredictably, eg from run-off entering a fill or from leakage of a water main, deliberate controlled inundation prior to building development may be a viable option or component of the ground improvement. Charles and Watts (1996) provide guidance and put forward a methodology for the assessment of collapse potential of fills. Their proposed field infiltration test, they suggest, needs further validation in a variety of fill types.

5.3.6 Controls

There is a degree of control in the flood trench method; eg it may be possible to have closer or deeper trenches. Little control appears possible on the densification process itself, however, as it relies on the passage of water through the fill and the breakdown of the lumps of fill. This is also the case when the source of the inundation is a rise in groundwater level. Collapse settlement is sudden and unpredictable and it is never really known when it will be completed.

5.3.7 References

Charles, J A, Earle, E W and Burford, D (1979)
"Treatment and subsequent performance of cohesive fill left by opencast ironstone mining at Snatchill experimental housing site, Corby"
Proc Conf on Clay Fills, London, 1978, Instn Civ Engrs, London, pp 63–72

Charles, J A, Hughes, D B and Burford, D (1985)
"The effect of a rise of water table in the settlement of backfill at Horsley opencast coal mining site, 1973 to 1983"
In: *Ground movements and structures* (J D Geddes, ed)
Pentech Press, London, p 198

Charles, J A and Watts, K S (1996)
"The assessment of the collapse potential of fills and its significance for building on fill"
Proc Instn Civ Engrs Geotechnical Engineering, Vol 119, Issue 1, Jan, pp 15–28

Clayton, C R I (1980)
"The collapse of a compacted chalk fill"
Proc Colloque Int sur le Compactage, Paris
Vol 1, pp 119–124

Davies, R V and Henkel, D J (1980)
"Geotechnical problems associated with the construction of Chater Station, Hong Kong"
Proc Conf on Mass Transportation in Asia, Hong Kong, Paper J3, pp 1–31

Donchev, P (1980)
Compaction of loess by saturation and explosion
Proc Colloque Int sur le Compactage, Paris
Vol 1, pp 313–317

Humpheson, C, Fitzpatrick, A J and Anderson, J M D (1986)
"The basements and substructure for the new headquarters of the Hong Kong and Shanghai Banking Corporation, Hong Kong"
Proc Instn Civ Engrs, Vol 80, PEI, August 1986, pp 851–883

Jennings, J E and Knight, K (1957)
"The additional settlement of foundations due to a collapse of structure of sandy subsoils on wetting"
Proc 4th Int Conf Soil Mech and Found Engg, Vol 1, pp 31–36

Litvinov, I M (1973)
"Deep compaction of soils with the aim of considerably increasing their carrying capacity"
Proc 8th Int Conf Soil Mech and Found Engg, Moscow, Vol 4.3, pp 365–367

Minkov, M, Evstatiev, D, Donchev, P and Steffanoff, G (1981)
"Compaction and stabilization of loess in Bulgaria"
Proc 10th Int Conf Soil Mech and Found Engg, Stockholm
AA Balkema, Rotterdam, Vol 3, pp 745–

Mitchell, J K. (1976)
Fundamentals of soil behaviour
J Wiley, New York

Mitchell, J K (1982)
"Engineering properties of tropical residual soils"
Proc Conf on Engineering and construction in tropical and residual soils, Honolulu
January, pp 30–57

Smyth-Osbourne, K R and Mizon, D J (1985)
"Settlement of a factory on opencast backfill"
In: *Ground movements and structures* (J D Geddes, ed)
Pentech Press, London, Vol 3, pp 463–479

Stroud, M A and Mitchell, J M (1990)
"Collapse settlement of old chalk fill at Brighton"
In: *Chalk*, Proc Int Chalk Symp, Brighton, September, 1989
Thomas Telford, London, pp 343–350

5.4 VACUUM PRE-LOADING

5.4.1 Definition

Vacuum pre-loading is the application of atmospheric pressure to form a temporary surcharge for soft clays while applying a vacuum to the surface of the soil beneath a membrane. The technique was developed in Sweden in the early 1950s.

5.4.2 Principle

Using atmospheric pressure, total stresses remain constant, while effective stresses are increased by the application of the vacuum so draining water from the ground towards the surface zone of reduced air pressure. Water is sucked out of the clay which can only consolidate (Kjellman, 1952). Figure 5.14 illustrates the principle and the increase in effective stresses.

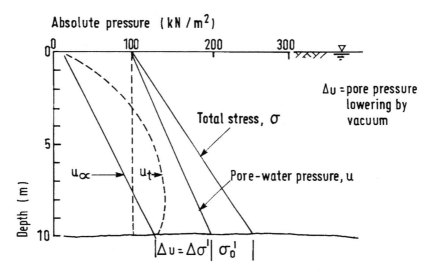

Figure 5.14   *Principle of vacuum pre-loading (after Choa, 1989)*

5.4.3 Description

An impermeable membrane is placed over a sand or gravel filter layer 150–500 mm thick, and sealed into the clay below the water table at its edges. A vacuum pump is connected, and air is pumped out of the porous filter and groundwater out of the underlying clay. Pressure differences of 0.6–0.8 atmospheres (60–80 kN/m^2) are the practical limit. This is equivalent to about 4–5 m of loose sand fill. Kjellman intended the method to be used with band drains, and Figure 5.15 illustrates the method.

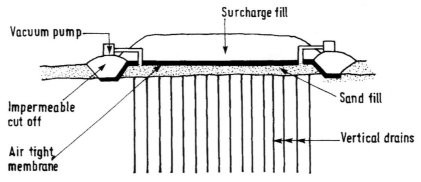

Figure 5.15   *Arrangement of vacuum pre-loading with surcharge and vertical drains (after Choa, 1989)*

Holtz and Wager (1975) note that vacuum pumps with a 500 mm-thick filter can easily maintain a vacuum of 60–80 kN/m^2 continuously for an area of about 5000 m^2.

To protect the pump the vacuum line should be brought into a tank or separator chamber to collect any water or gas, which can be drawn off periodically. Pumping can be switched on at, say, 60 kN/m^2, and off at 80 kN/m^2, to reduce pumping costs. An alternative to a membrane is to use a series of pumping wells (Pilot, 1978). A variation of these techniques was reported in *Ground Engineering* (Anon, 1998) in which narrow (0.25 m), deep trenches (as much as 7 m deep) are filled with sand and dewatered. The trenches are sealed at ground level so that as the dewatering continues a vacuum is generated.

5.4.4   Applications and limitations

It is not known if vacuum pre-loading has been used in UK, but there are case histories from several European countries, from North America and the Far East. Two examples are summarised in Table 5.4. The idea has been taken up in China and Russia (Ye *et al*, 1983; Chen and Bao, 1983; and Arutinian, 1983). In China, water is ponded to form a seal for the membrane over the gravel.

Table 5.4   *Two examples of vacuum pre-loading (after Pilot, 1978)*

Site (reference)	Ground conditions	Method	Settlement (mm)
Inland Sea (Japan) – reservoir foundation	6 m of clay and silt overlying sands and gravels	Well method (ie without membrane) pumping from well points and sand drains	300–600 after 60 days' application
Brest (France) – hydraulic fill	6 m of clayey silt fill overlying 2 m of mud above sands and gravels	Well method (ie without membrane) 20–45 kN/m^2 negative pressure – well spacing 10 m – well spacing 3.3 m	after about 30 days 30–120 190–420

Vacuum pre-loading might well be a promising alternative to the use of filling as a temporary surcharge, or it could be used in conjunction with a lesser height of fill. Lateral stability problems can be minimised with vacuum pre-loading (Hansbo, 1979), although the process should be applied to relatively homogeneous rather than stratified ground. Permeable layers tend to reduce the effectiveness of the vacuum.

More recently, case histories have been reported by Choa (1989), Woo *et al* (1989), and Sehested and Yee (1990). Practical problems were experienced by Sehested and Yee in the effectiveness and serviceability of their vacuum pumps. Placing fill on the membrane improved the seal. Surface roots, earthworms and birds also caused problems by creating holes in the sealing membrane. Nevertheless, in these case histories vacuum pre-loading compared well with surcharge loading.

Halton *et al* (1961) describe a variation of the vacuum pre-loading idea. About 6 m of organic silts and clays formed the seal above a deep granular layer. The vacuum was then created using vacuum pumps and combined with the use of deep wells, in conjunction with sand drains.

5.4.5 Design

Field trials are essential to establish how to achieve a good seal with the membrane and to ensure that the vacuum pumps work efficiently.

The principle of increasing the effective stress by applying a vacuum is straightforward, but reference should be made to Kjellman (1952) and subsequent case histories for the practical details.

5.4.6 Controls

The only real control is to check that the vacuum is maintained over the full area of treatment and that it affects sufficient depth of soil. Well (or trench) spacing and depth are factors that can be changed for different conditions. The progression of settlement with time is the main observation of treatment effectiveness. This can be quite rapid, ie of the order of weeks.

5.4.7 References

Anon (1988)
"Vacuum packed"
Ground Engineering, Vol 31, No 2, February, pp 18–19

Arutinian, R N (1983)
"Vacuum-accelerated stabilisation of soils in landslide body"
Proc 8th Eur Conf Soil Mech and Found Engg, Helsinki
AA Balkema, Rotterdam, Vol 2, pp 575–578

Chen, H and Bao, X-C (1983)
"Analysis of soil consolidation stress under the action of negative pressure"
Proc 8th Eur Conf Soil Mech and Found Engg, Helsinki
AA Balkema, Rotterdam, Vol 2, pp 591–596

Choa, V (1989)
"Drains and vacuum preloading pilot test"
Proc 12th Int Conf Soil Mech and Found Engg, Rio de Janeiro
Vol 2, pp 1347–1350
Ground Engineering (1998)

Halton, G R, Loughney, R W and Winter, E (1961)
"Vacuum stabilisation of subsoil beneath runway extension at Philadelphia International Airport"
Proc 5th Int Conf Soil Mech and Found Engg, Montreal

Hansbo, S (1979)
"Consolidation of clay by band-shaped prefabricated drains"
Ground Engineering, Vol 12, No 5, July, pp 16–25

Holtz, R D and Wager, O (1975)
"Preloading by vacuum: current prospects"
In: *Soil and rock mechanics, culverts and compaction*
Transportation Research Record 548, Washington DC, pp 26–29

Kjellman, W (1952)
"Consolidation of clay soil by means of atmospheric pressure"
Proc Conf on soil stabilisation, MIT, Cambridge, Mass, pp 258–263

Pilot, G (1978)
"State of the art"
Bulletin de Liaison des Laboratoires des Ponts et Chaussées, Soil Mechanics,
Special Issue VI, April, pp 140–178

Sehested, K G and Yee, T S (1990)
"Soil improvement using vertical band drains and vacuum preloading at Section 6/7"
Proc Int Symp on Trial Embankments on Malaysia Marine Clays, Kuala Lumpur
Vol 24, pp 2/89-101, (R R Hudson, C T Toh and S F Chan, eds), Malaysian Highway
Authority

Woo, S M, Van Weele, A F, Chottivittayathanin, R and Trangkarahart, T (1989)
"Preconsolidation of soft Bangkok Clay by vacuum loading combined with non-
displacement sand drains"
Proc 12th Int Conf Soil Mech and Found Engg, Rio de Janeiro, Vol 2, pp 1431–1434

Ye, B-R, Lu, S-Y and Tang, Y-S (1983)
"Packed sand drain: atmospheric pre-loading for strengthening soft foundations"
Proc 8th Eur Conf Soil Mech and Found Engg, Helsinki
AA Balkema, Rotterdam, Vol 2, pp 717–720

5.5 DEWATERING FINE-GRAINED SOILS

5.5.1 Definitions

Dewatering is the removal of pore water from the ground. This may be achieved in
permeable deposits by pumping from bored wells or by the installation of wellpoints,
eductor (or ejector) systems, and, in fine-grained soils by vacuum or by electro-osmosis.

The techniques of groundwater control by pumped wells, wellpoints and eductors are
frequently used in temporary works and are described by Preene *et al* (1997), Somerville
(1986) and Miller (1988). Eductors are used for finer-grained soils in which wells or
wellpoints would not be effective, and they operate by applying a vacuum to the
surrounding ground by a high pressure water Venturi system. As the main purpose of
these dewatering techniques is control of groundwater rather than ground improvement
they are not described here. The reader is referred to Preene *et al* (2000), Powrie and
Preeene (1994) and Preene and Powrie (1994) for further information.

The application of a vacuum to deep wells is to enhance their yield. Occasionally this
may be used for pumped wells in confined aquifers; for ground improvement the
technique is a variation of vacuum pre-loading (Section 5.4) or an addition to vertical
drains (Section 5.2).

Dewatering by electro-osmosis is the removal of pore water from fine-grained soils by
the application of an electrical potential difference, causing the water to flow to the
cathode well.

Figure 5.16 shows the application ranges in terms of coefficient of permeability, which
are usually suitable for different dewatering methods.

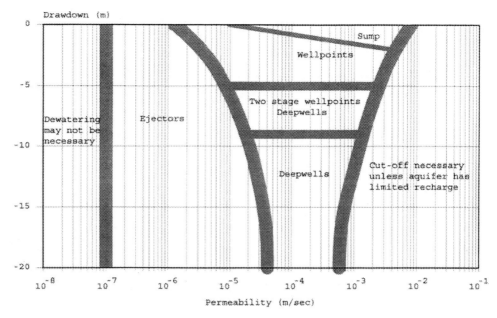

Figure 5.16 *Application range for different methods of dewatering (after Roberts and Preene, 1993)*

5.5.2 Principle

The removal of pore water increases the effective stress in the ground while the total stress remains unchanged. Primary consolidation settlement rates are increased, and there may be consequential improvement in the ground's shear strength, which would benefit the stability of slopes and excavations.

5.5.3 Deep wells with vacuum

A vacuum can be applied to deep wells to increase their yield. Powers (1981) shows an example of how the well is sealed. This technique derives from abstraction of groundwater for water supply, particularly from confined aquifers. The pumping/vacuum system has to be designed for long service.

Deep wells appear to be used rarely for ground improvement. When they are employed, it is usually in conjunction with band drains (Wolski *et al*, 1979) or with a surcharge (Tomlinson and Wilson, 1973). Steger (1987) reports the use of dewatering alone as being sufficient for the improvement of loose sand over cap-rock for a water pollution plant in Bahrain. Deep wells are probably not very suitable as a means of ground improvement because their use can cause regional settlements (Hausser *et al*, 1964; Steger, 1970). These can damage adjoining structures and utilities.

An attempt was made to use deep wells in conjunction with surcharge for construction of a trial embankment in Malaysia. Unfortunately, the bank failed at about 4 m height before reaching the design height of 6 m (Hudson *et al*, 1990). Greenwood (1988) describes one failure and one success with vacuum wells.

Guidance on the design of deep wells and pumping equipment is given in Powers (1981). Once again field trials appear to be essential, to examine design assumptions and develop operational procedures.

5.5.4   Electro-osmosis

Description

Electro-osmosis was developed in Germany in the 1930s. It involves the forced movement of pore water from one electrode to another where an electrical potential is applied across a saturated porous material. If the negative electrodes (cathodes) are designed as drainage wells then the porewater will be compelled to flow to the wells, from where it can be removed (Casagrande, 1949). Reversing a naturally existing flow of water can also be used to stabilise unstable silt slopes or "quicksand" conditions.

The electrolytic medium is the flowing pore water. The reduction in pore water pressure increases shear strength; at the same time the pH value of the ground also increases. Farmer (1975) advises that electro-osmosis is a decelerating process in all soils, however. As pore water content decreases the electrolyte concentration increases, with flow being further retarded by gas generation at the electrodes. Reversing polarity at this point, however, can increase consolidation pressure and give a more uniform improvement in strength and decrease in water content (Wan and Mitchell, 1976).

Applications and limitations

The primary references to electro-osmosis are those of Casagrande, who developed the technique (see Casagrande, 1983, for earlier references, and Gray and Mitchell, 1967). Most examples of its successful use are for the stabilisation of silty clays and silts.

Cathode wells can be constructed in several ways. Essentially, they combine metal rods or casing as the cathode and well screens (incorporating suitable filters for silt and clay soils) to permit water entry and to contain an eductor pump. The anodes are either reinforcing bars or aluminium rods. The eductors are connected by a header main to the high-pressure water supply that operates the Venturi suction to draw water from the well.

An example of a use of electro-osmosis to stabilise an excavation for a railway cutting is given in Figure 5.17. Wilkins and Chandler (1990) list 33 case histories from around the world; to this list can be added Casagrande *et al* (1982) and Eggestad and Føyn (1984).

The potential difference to be applied and for how long depends on the soil type and the degree of improvement required. Case histories give values ranging from 6 volts/metre at 4 amps for 15 days to 65 V/m at 250 A for four months. Wilkins and Chandler (1990) also report an unsuccessful use of electro-osmosis in Malaysian soft marine clays. They note that if dissolved salts are in the range of 6000–14 000 parts per million, then electro-osmosis may be ineffective.

Direct electrical current can also be used to move solutions through the ground. Mitchell (1982) describes this process as *electro-chemical injection* and suggests that it can be applied to silty soils instead of grouting. Chemical stabilisers are introduced at the anode and are carried towards the cathode using electrical gradients of 50–100 V/m. Broms (1979) advises that electro-chemical stabilisation should not be used when the conductivity of the soil is high. The method is considered to be expensive and should only be employed in special cases.

Design

Some guidance on rate of water flows is given by Wilkins and Chandler (1990). Casagrande (1983) prepared a state-of-the-art report on the technique, which is a key reference. Field trials would be essential.

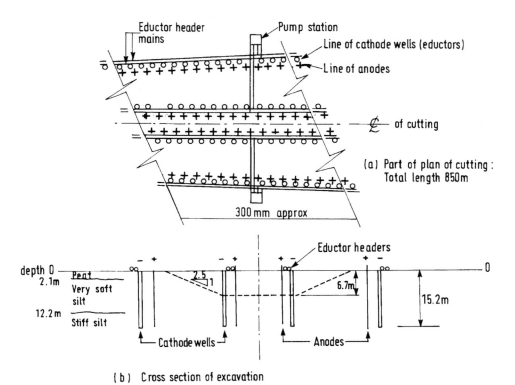

Figure 5.17   *Layout of an electro-osmosis system (after Casagrande et al, 1982)*

5.5.5   References

Broms, B B (1979)
"Problems and solutions to construction in soft clays"
Proc 6th Reg Conf Soil Mech and Found Engg, Singapore, Vol 2, pp 27–36

Casagrande, L (1949)
"Electro-osmosis in soils"
Géotechnique, Vol 1, No 3, pp 159–177

Casagrande, L (1983)
"Stabilization of soils by means of electro-osmosis – state of the art"
J Boston Soc Civ Engg, Am Soc Civ Engrs, Vol 69, No 2, pp 255–302

Casagrande, L, Wade, N, Wakely, M and Loughney, R (1982)
"Electro-osmosis projects in British Columbia, Canada"
Proc 10th Int Conf Soil Mech and Found Engg, Stockholm, 1981
AA Balkema, Rotterdam, Vol 3, pp 607–610

Eggestad, A and Føyn, T (1984)
"Electro-osmotic improvement of a soft sensitive clay"
Proc 8th Eur Conf Soil Mech and Found Engg, Helsinki, 1983
AA Balkema, Rotterdam, Vol 2, pp 597–603

Farmer, I W (1975)
"Electro-osmosis and electro-chemical stabilization"
Chapter 3 in: *Methods of treatment of unstable ground* (R G Bell, ed)
Newnes-Butterworth, London

Gray, D H and Mitchell, J K (1967)
"Fundamental aspects of electro-osmosis in soils"
J Soil Mech and Found Div, Am Soc Civ Engrs, Vol 93, SM6, pp 209–236

Greenwood, D A (1988)
"Substructure techniques for excavation support"
Proc Conf on Economic Construction Techniques
Instn Civ Engrs, London, November

Hausser, P C G, Finlinson, J C H and Elliott, J A (1964)
"A comparison of the design and construction of dry docks at Immingham and Jarrow"
Proc Instn Civ Engrs, Vol 27, February, pp 291–324

Hudson, R R, Toh, C T and Chan, S F (eds) (1990)
Trial Embankments in Malaysian Marine Clays
Malaysian Highway Authority

Miller, E (1988)
"The eductor dewatering system"
Ground Engineering, Vol 21, No 6, September, pp 29–31

Mitchell, J K (1982)
"Soil improvement – state of the art"
Proc 10th Int Conf Soil Mech and Found Engg, Stockholm, 1981
AA Balkema, Rotterdam, Vol 4, pp 509–565

Preene, M and Powrie, W (1994)
"Construction dewatering in low permeability soils: some problems and solutions"
Proc Instn Civ Engrs Geotechnical Engineering, Vol 107, January, pp 17–26

Powers, J P (1981)
Construction dewatering: a guide to theory and practice
J Wiley, New York

Powrie, W and Preene, M (1994)
"Performance of ejectors in construction dewatering systems"
Proc Instn Civ Engrs Geotechnical Engineering, Vol 107, July, pp 143–154

Preene, M, Roberts, T O L, Powrie, W and Dyer, M R (2000)
Groundwater control: design and practices
Publication C515, CIRIA, London

Roberts, T O L. and Preene, M (1993)
"Range of application of construction dewatering systems"
Int Conf on Groundwater problems in Urban Areas
Instn Civ Engrs, London, June

Somerville, S H (1986)
Control of groundwater for temporary works
Report 113, CIRIA, London

Steger, E H (1970)
Discussion in *Proc Conf on Ground Engineering*, June
Instn Civ Engrs, London, p 81

Steger, E H (1987)
"Water pollution control centre, Tubli, Bahrein"
In: *Building on marginal and derelict land*
Thomas Telford, London, pp 69–81

Tomlinson, M J and Wilson, S M (1973)
"Preloading of foundations by surcharge on filled ground"
Géotechnique, Vol 23, No 1, March, pp 117–124

Wan, T Y, and Mitchell, J K (1976)
"Electro-osmotic consolidation of soils"
J Geotech Engg, Am Soc Civ Engrs, Vol 102, GT5, May, pp 473–491

Wolski, W, Furstenberg, A and Lechowicz, Z (1979)
"Consolidation parameters of a soft clay layer improved by means of vertical drains"
In: *Design parameters in geotechnical engineering*
Proc 7th Eur Conf Soil Mech and Found Engg, Brighton, 1979
British Geotech Soc, London, Vol 3, pp 311–314

Wilkins, E A and Chandler, B C (1990)
"Electro-osmosis trial, Section 6/3, Muar Flat"
Proc Int Symp on Trial Embankments on Malaysian Marine Clays (R R Hudson, C T Toh and S F Chan, eds), Kuala Lumpur, Malaysian Highway Authority, Vol 2, pp 2/48–52

5.6 PRESSURE BERMS

5.6.1 Definition

Pressure berms are fills extending the side slopes of embankments on soft clays to increase bearing capacity and to prevent rotational failures at the toe of the bank itself.

5.6.2 Principle

The additional weight and extent of the berms increase the normal stresses on any potential failure surface. The added load beyond the embankment also induces added settlement, but with pore-water pressure dissipation there is usually a consequential increase in shear strength of the ground.

5.6.3 Description

Figure 5.18 shows two forms of pressure berm. The fill should be placed all the way across the section, although there have been occasions when the berms have been placed later as a stability measure. Several horizontal berms may be needed to stabilise a high embankment, while sloping berms may be used when undrained shear strength increases with depth (Broms, 1979).

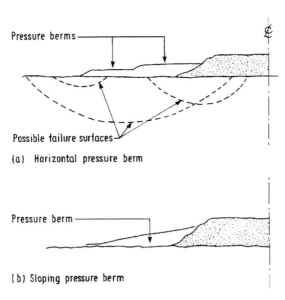

Figure 5.18   *Pressure berms*

5.6.4 Applications and limitations

Figure 5.18 illustrates the main disadvantage of pressure berms in that a far greater area is needed. It is also likely that total settlements will be more than if the embankment had been constructed without berms. Pressure berms (when laid as clay blankets) can be used to control the seepage forces on both the landward and river sides of dams or levées.

5.6.5 Design

Stability may have to be assessed using a wedge analysis where a weak layer overlies strong ground; alternatively a circular failure surface can be assumed for the assessment of the placing of each layer of filling as a form of toe loading. The effect of a desiccated crust can be important, as it will increase bearing capacity.

Hudson *et al* (1990) give a typical example of a pressure berm for a trial embankment 6 m high on very soft Malaysian marine clays. Its behaviour was compared with eight other embankments planned to be of comparable height, and using various forms of ground improvement. Broms (1979) gives guidance on the design of pressure berms.

5.6.6 Controls

These are likely to be essentially the same as those required for constructing embankments on soft soils and could involve controls over the height and frequency of lifts and the measurement of pore pressures and lateral displacements.

5.6.7 References

Broms, B B. (1979)
"Problems and solutions to construction in soft clay"
Proc 6th Asian Reg Conf Soil Mech and Found Engg, Singapore
Vol 2, pp 35–38

Hudson, R R, Toh, C T and Chan, S F (eds) (1990)
Proc Int Symp on Trial Embankments on Malaysia Marine Clays
Kuala Lumpur, Malaysia Highway Authority

6 Improvement by structural reinforcement

Many ground improvement techniques could be considered as a form of reinforcement. Stone columns, for example, are introduced materials that stiffen the ground; some grouts strengthen the mass of soil into which they are injected. The distinction drawn for the classification used in this report of structural reinforcement is that prefabricated tensile or shear elements are installed in the ground with the purpose of forming a composite material.

The still-developing techniques of reinforced soil, where earth-fill structures are created by the inclusion of metallic, polymeric or geotextile elements, are described in Section 6.1.

The insertion of rod reinforcement into a cut face of soil to improve its stability, soil nailing, is described in Section 6.2.

Section 6.3 explains the uses of root- or micro-piles to form units of strengthened soil for different engineering purposes.

Two other types of structural reinforcement are also described, slope dowels (in Section 6.4) where additional shear resistance is provided to potentially or actually unstable slopes, and the use of piles in compressible soils below embankments (Section 6.5).

Friction pile groups are perhaps the commonest form of structural reinforcement, a combination of soil and the introduced piles, which rely on the adhesion or friction of the soil for their carrying capacity to achieve a much greater bearing strength than the soil alone. Bearing piles of this conventional form are not included in this report, nor are ground anchorages and rock bolts, as these are fully described elsewhere (eg Wynne, 1988, for bearing piles; Douglas and Arthur, 1983, for rock reinforcement; and BS 8081:1989 for ground anchorages).

Soil reinforcement is not new: forms of geotextiles have been used for thousands of years all over the world (Giroud, 1986). Jones (1996) pointed out an example given in the Bible; and Yamanouchi (1986) explained how bamboo reinforcement, particularly in waterfront structures and for improvement of the ground surface, was used hundreds of years ago in Japan and China.

From sandbags for sea defences in the 1950s the applications for geotextiles have rapidly increased (Giroud, 1986). In the early 1960s, Vidal in France developed the technique called reinforced earth (for much of the time since using metallic strip reinforcing elements). The development and use of synthetic fibres in the textile and carpet industries and need for a large-use market led to their use in civil engineering. Synthetic materials are used for reinforcing, filtration and drainage, and separation and confinement.

The functions of the structural reinforcement are listed in Table 6.1 and their applications shown in Table 6.2. Note that combinations of methods are possible.

Table 6.1   *Types of improvement by structural reinforcement (after Schlosser and Juran, 1981)*

Function	Improvement method				
	Reinforced soil	Soil nailing	Micro-piles	Slope dowels	Embank-ment piles
Tension	✓✓✓	✓✓✓	✓		
Compression			✓✓✓		✓✓✓
Shear		✓		✓✓✓	✓
Bending			✓	✓	

Key: ✓✓✓ main function   ✓ subsidiary function

Table 6.2   *Application of structural reinforcement methods (after Schlosser and Juran, 1981)*

Application	Improvement method				
	Reinforced soil	Soil nailing	Micro-piles	Slope dowels	Embank-ment piles
Bearing capacity	✓		✓✓✓		✓✓✓
Stability	✓✓✓	✓✓✓	✓	✓✓✓	
Settlement management	✓				✓✓✓

Key: ✓✓✓ Main application   Subsidiary application

Note that references cited in this introduction are given at the end of Section 6.1.

6.1   REINFORCED SOIL

6.1.1   Definition

Reinforced soil is the combination of compacted earth fill with tensile reinforcement elements to create an earth structure, whose properties and performance depend on the interaction between the soil and the reinforcement. The reinforcement is laid between layers of compacted earth fill, so being an intrinsic part of the construction process. The reinforcing elements may be metallic, polymeric or even natural materials, but usually they are prefabricated in the form of strips, grids, meshes, webbing, nets or fabric sheets.

The trade names *Reinforced Earth (Terre Armée)* and *Anchored Earth* are particular forms of reinforced soil.

6.1.2   Principle

The principle of reinforced soil is that an introduced material provides a tensile restraining force that reduces the lateral stress required to maintain the equilibrium of a loaded soil unit. This is illustrated in Figure 6.1. As the soil element compresses under vertical stress and tends to strain laterally, a tensile stress is generated in the reinforcement, resisting the outward movement and giving rise to lower horizontal stresses than the same soil element under the same vertical load but without reinforcement. It is important to note that the tensile force in the reinforcing element depends on there being lateral strain.

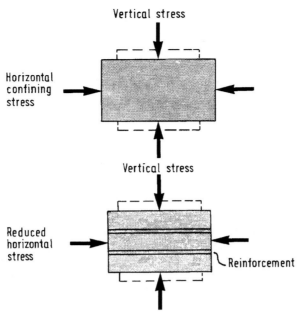

Figure 6.1   *The principle of reinforced soil*

6.1.3   Description

All reinforced soil structures are combinations of suitable earth fill usually with several layers of the reinforcing elements placed on compacted fill. The technique is used to construct: (1) vertical walls and abutments, when facing panels are used in conjunction with the reinforcing elements; (2) slopes of embankments steeper than would be stable (or at an acceptable degree of stability) with unreinforced soil; (3) embankments on soft soils, where the foundation soil has inadequate bearing capacity to support the height of fill; (4) unpaved roads; (5) special mattress-type foundations or stress-reducers for soft or backfilled ground; and (6) repairs to slipped material of earth slopes. The first four of these forms of reinforced soil structure are illustrated conceptually in Figure 6.2.

The great majority of structures built of reinforced soil use the *Reinforced Earth* method of Vidal. Walls have been built up to 34 m high (Bonaparte *et al*, 1989). Worldwide, more than 10 000 structures incorporating 6 000 000 m^2 of facing have been built (BSI, 1991) for retaining walls, sea walls and quays, bridge abutments, blast walls etc.

The *Reinforced Earth* method combines metallic strips and compacted granular fill. Prefabricated facing elements (Figure 6.3) are connected to the metal strips as the fill is raised, primarily as surface protection and soil containment. Although giving external form to the soil structure and containing the fill, the facing units do not themselves provide structural restraint to the full lateral stresses from the fill. Depending on the situation and the aesthetic requirements of the structure, the facings may be galvanised steel or pre-cast concrete panels or blocks.

Many other types of reinforcement are now available, from sheets of strong geotextile fabrics to strips, grids and meshes of polymers (see Figure 6.4). For a short-term use the fabric can be wrapped around the fill to contain it as it is raised (Figure 6.5), though it may be necessary to use temporary shuttering. With the prefabricated polymeric strips and grids, pre-cast concrete facing panels are used, although various other materials are possible – timber, stone-filed gabions, and hydroseeded grassing are examples.

In reinforced soil walls and steep slopes, the reinforcing layers are usually at vertical spacings of 300–750 mm, with which the (one or more) compaction lifts of the fill would have to be consistent. The fill specification is important, usually being limited to a well-graded granular material, although clay, chalk and some industrial spoil fills may be used in clearly defined circumstances (see Temporal *et al*, 1989, for a study of the use of chalk and clay fills win soil reinforcement).

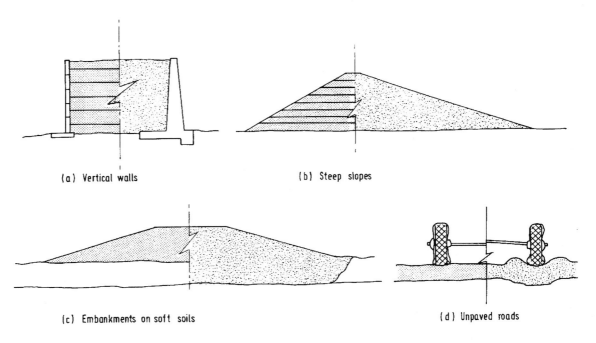

(a) Vertical walls   (b) Steep slopes

(c) Embankments on soft soils   (d) Unpaved roads

Figure 6.2   *Concepts of reinforced soil*

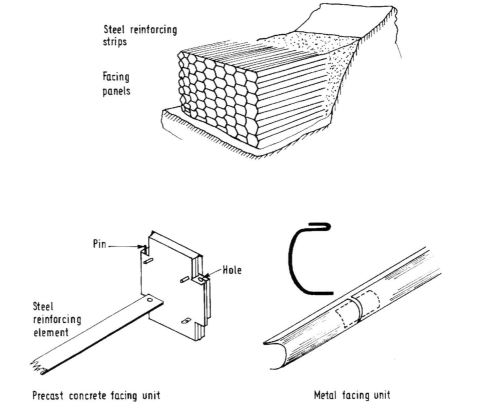

Steel reinforcing strips

Facing panels

Pin

Hole

Steel reinforcing element

Precast concrete facing unit   Metal facing unit

Figure 6.3   *Reinforced earth*

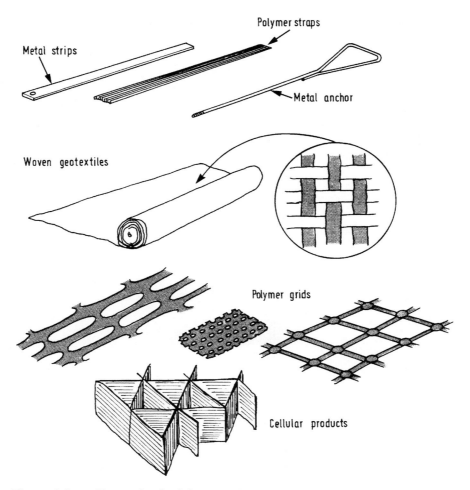

Figure 6.4   *Types of soil reinforcement*

For the construction of embankments on soft soils the reinforcement can be laid directly on the soft foundation, where it provides a working platform for fill placement (Figure 6.6). It may be a strong woven geotextile, a geo-grid (see Figure 6.4) or a cellular mattress filled with granular material (Figures 6.4 and 6.7).

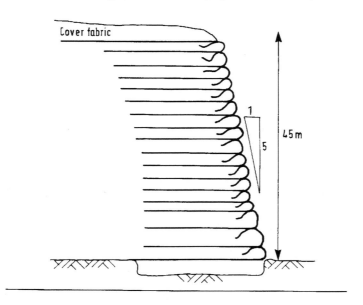

Figure 6.5   *Wrap-around geotextile wall*

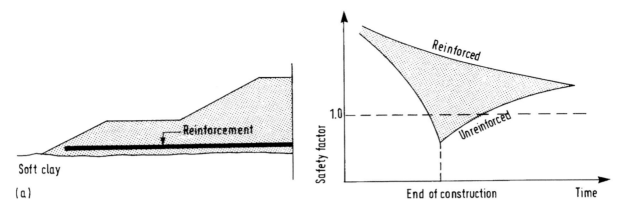

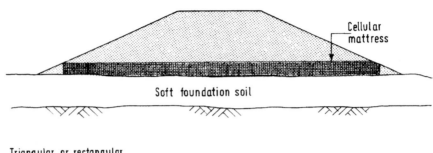

Figure 6.6 *Embankment on soft soil: (a) reinforcement at base, (b) safety factor and time for reinforced and unreinforced cases (after Jewell, 1996)*

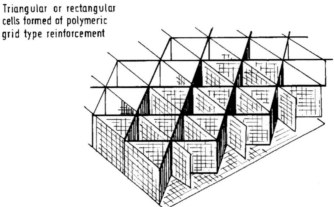

Figure 6.7 *Cellular mattress*

The use of reinforcement for slip repairs is illustrated in Figure 6.8. The slipped soil is removed and can be replaced (lime can be added to clays to enable plant to work on them more easily) and recompacted, or granular fill may be imported.

Developments in this rapidly changing series of techniques include the use of waste products such as rubber tyres, either as facing or anchorage elements, and random reinforcement mixed into the soil. This can take the form of a continuous filament of glass fibre (Leflaive, 1985) or small square geogrid mesh elements (Mercer *et al*, 1985; McGown *et al*, 1987).

Anchored earth (Figure 6.9) is an alternative for wall and abutment construction. Here the resistance is mobilised principally by the development of bearing pressure on an anchor at the end of a straight shaft (Murray and Irwin, 1982; Jones *et al*, 1985). Geogrids and geotextiles may also be used for isolated foundations, for paved and unpaved roads or large loaded areas over soft ground.

In all these uses where fill is placed on soft ground, the sheet of reinforcing material may act as a separator, preventing contamination of the fill by fines of the underlying soil or penetration of the fill into the soft soil below. This useful, important function is also performed by low-strength geotextile fabrics, such as non-woven ones, although they would not be suitable where reinforcement is the primary requirement.

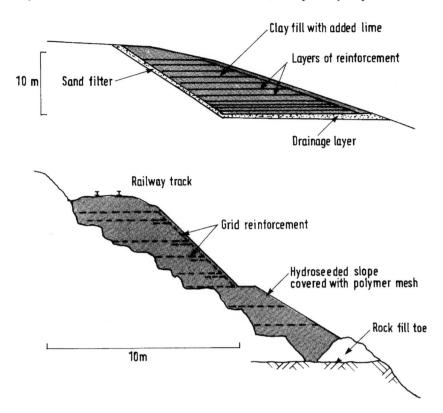

Figure 6.8 *Reinforced soil for slip repairs: slip repair to motorway cutting (top) and railway embankment (bottom)*

6.1.4 Applications

The range of applications of reinforced soil covers virtually all earth-fill structures from unpaved roads to embankments, from temporary vertical walls to permanent bridge abutments. It also includes waterfront walls. The advantages are several:

- smaller quantities of earth fill are needed

- steeper embankment slopes reduce the land take required

- construction can be direct on soft ground

- the technique is not a separate operation in the construction process, but is integrated into the placing and compaction of the earth fill: structures can often be built more quickly than by conventional methods.

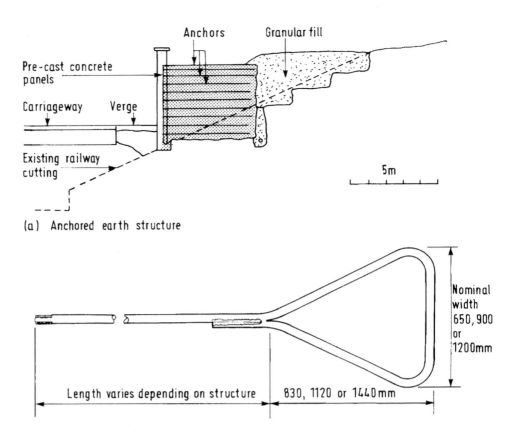

Anchors   Granular fill

Pre-cast concrete panels

Carriageway   Verge

Existing railway cutting

5m

(a) Anchored earth structure

Nominal width 650, 900 or 1200mm

Length varies depending on structure | 830, 1120 or 1440mm

(b) Anchor shape in plan

Figure 6.9   *Anchored earth (after Snowdon, Darley and Barratt, 1986)*

There is increasing confidence in the long-term performance of the materials from which the reinforcing elements are fabricated so that they are not used only for temporary works or those expected to have a short life. Nevertheless, uncertainty remains about the durability of some materials and the conditions they will experience.

6.1.5   Limitations

Reinforced soil relies upon deformation for its effectiveness, ie soil strain has to be transferred to the reinforcement for it to develop its tensile or bearing resistance. The advantage is that reinforced soil structures tolerate differential settlements, and so have been used in areas of mining subsidence. The facing units can be affected, however. Schlosser and Guilloux (1985) recommended that differential settlements should not exceed 1 or 2 per cent for pre-cast concrete facings or 3 per cent for metallic facings. Constructing reinforced soil walls requires very careful control of line and level and co-ordination with the compaction of fill layers.

Some doubt remains about the durability and long-term performance of the reinforcing material and components of the facing system. Severe and rapid corrosion of steel reinforcement, even when galvanised, is possible (Blight and Dane, 1989). The extrapolation of polymer properties over very long times and at possibly high tempera-tures under sustained load is still being studied. There are even more unknowns about their behaviour when confined in soil. Polymeric materials degrade when exposed to ultra-violet rays and can be damaged by rough handling on site or by sharp stones in fill.

When constructing on soft soils care has to be taken not to overstress the reinforcement: the tensile forces induced by the filling operations may be greater than those required to give stability (BSI, 1995). Even with light small bulldozers spreading the first layer of fill, a geotextile will deform considerably and mudwaves of the underlying soft clay be generated ahead and to the side of the filling (Figure 6.10). For multi-reinforced earth structures, and particularly those of long design lives, the design details should take account of erosion protection and drainage. Reinforcement that relies upon friction, eg strips and fabric sheets, will be less effective in fills containing too much silt and clay. The fines content of the granular backfill is usually limited to 15 per cent.

6.1.6   Design

Design methods for reinforced soil are given in BSI (1995) *Code of practice for strengthened/reinforced soil and other fills*. The Department of Transport Advice Note HA68/94 sets out design procedures for steepened slopes using reinforced soil and soils nails. CIRIA Special Publication 123 (Jewell, 1996) provides detailed guidance on the use of geotextiles for soil reinforcement applications of vertical walls, steep slopes, unpaved roads, embankments on soft soils and for slip repairs.

For reinforced soil retaining walls, both external and internal stability have to be satisfied. For external stability, the wall should be checked conventionally against overturning, sliding, bearing capacity failure and overall stability. Sliding usually governs (Jones *et al*, 1987), except in landslide correction schemes where overall stability is likely to control geometry. Detailed guidance for walls is given by Jones (1996) and Jewell (1996). Jewell and Milligan (1989) provide charts of horizontal deflection and vertical settlement.

Figure 6.10   *Mudwaves on filling over a geotextile on soft clay*

For internal stability, the tension in the reinforcement and the pull-out or rupture resistance must be established. Tension resistance of the reinforcement depends greatly on the lateral earth pressure coefficient. This in turn depends on the degree of restraint created by the reinforcement elements, the amount of compaction, and vertical soil stress. Pull-out resistance of the reinforcing elements is provided by friction, and by transverse bearing resistance. One mechanism will dominate depending on the type of reinforcement that is used (Jones *et al*, 1987).

Reinforced soil slopes can be designed in a similar manner to walls using the method of Jewell *et al* (1985). The design approach for the reinstatement of slopes is given by Murray (1985), and that for the reinforcement of embankments by Milligan and La Rochelle (1985). Tensile reinforcement can also be applied beneficially to the granular fill for unpaved roads. A design approach is described by Milligan *et al* (1989).

6.1.7 Controls

Construction controls relate to (1) the quality of the reinforcement and facing materials in manufacture and their proper handling and storage on site; (2) to the workmanship in accurate, reliable assembly while the earthworks and facing operations also take place; and (3) the selection, placing and compaction of the imported fill.

Monitoring the reinforced earth structure (eg for verticality or settlement) may be necessary depending on its sensitivity. On soft foundations, the ground movements can be so great that usual forms of geotechnical instrumentation may not be able to cope; they could be rendered unusable because of the potentially large near-surface displacements. Alternatives could involve using simple slip indicators (ie to show if movement is taking place and at what depth) and surveying from well outside the working area.

6.1.8 References

Blight, G E and Dane, MS W (1989)
"Deterioration of a wall complex constructed of reinforced earth"
Géotechnique, Vol 39, No 1, March, pp 47–53

Bonaparte, R, Schmertmann, G R, Chu, D and Chovery-Curtis, V E (1989)
"Reinforced soil buttress to stabilise a high natural slope"
Proc 12th Int Conf Soil Mech and Found Engg, Rio de Janeiro, Vol 2, pp 1227–1230

BSI (1989)
BS 8081: 1989 *Code of practice for ground anchorages*
British Standards Institution, London

BSI (1995)
BS 8006: 1991 *Code of practice for strengthened/reinforced soils and other fills*
British Standards Institution, London

Douglas, T H and Arthur, L J (1983)
A guide to the use of rock reinforcement in underground excavations
Report 101, CIRIA, London

Giroud, J-P (1986)
"From geotextiles to geosynthetics: a revolution in geotechnical engineering"
Proc 3rd Int Conf on Geotextiles, Vienna, Vol 1, pp 1–18

Highways Agency (1994)
Design methods for reinforcement of highway slopes by reinforced soil and soil nailing techniques
Highways Agency Advice Note HA68.94, HMSO, London

Jewell, R A (1996)
Soil reinforcement with geotextiles
Special Publication 123, CIRIA, London/Thomas Telford, London

Jewell, R A and Milligan, G W E (1989)
"Deformation calculations for reinforced soil walls"
Proc 12th Int Conf on Soil Mech and Found Engg, Rio de Janeiro
AA Balkema, Rotterdam, Vol 2, pp 1257–1262

Jewell, R A, Paine, N and Woods, R I (1985)
"Design methods for steep reinforced embankments"
In: *Polymer grid reinforcement*, Thomas Telford, London, pp 70-89

Jones, C J F P (1996)
Earth reinforcement and soil structures
Butterworths, London, 2nd edn

Jones, C J F P, Murray, R T, Temporal, J and Mair, R J (1985)
"First application of anchored earth"
Proc 11th Int Conf Soil Mech and Found Engg, San Francisco, Vol 3, pp 1709–1712

Jones, C J F P, Cripwell, J P and Bush, D I (1987)
"Reinforced earth trial structure for Dewsbury Ring Road"
Proc Instn Civ Engrs, Vol 88, Part 1, April, pp 321–345

Leflaive, E (1985)
"Soil reinforcé par des fils continus: le Tesol"
Proc 11th Int Conf Soil Mech and Found Engg, San Francisco, Vol 3, pp 1787–1790

McGown, A, Andrawes, K Z, Hytiris, N and Mercer, F B (1987)
"Improvement of marginal soils by mixing in polymeric mesh elements"
In: *Building on marginal and derelict land*
Thomas Telford, London, pp 775–779

Mercer, F B, Andrawes, K Z, McGown, A and Hytiris, N (1985)
"A new method of soil stabilisation"
In: *Polymer grid reinforcement*
Thomas Telford, London, pp 244–249

Milligan, V and La Rochelle, P (1985)
"Design methods for embankments over weak soils"
In: *Polymer grid reinforcement*
Thomas Telford, London, pp 95–102

Milligan, G W E, Jewell, R A, Houlsby, G T and Bard, H J (1989)
"A new approach to the design of unpaved roads: Parts I and II"
Ground Engineering, Vol 22, No 3, pp 25.9 (Part I) and No 8, pp 37–42 (Part II)

Murray, R T (1985)
"Reinforcement techniques in repairing slope failures"
In: *Polymer grid reinforcement*
Thomas Telford, London, pp 47–53

Murray, R T and Irwin, M J (1982)
A preliminary study of TRRL Anchored Earth
Supplementary Report SR674, Transport and Road Research Laboratory, Crowthorne

Schlosser, F. and Guilloux, A. (1985)
"Reinforced earth uses on soft soils"
Proc Symp Recent developments in Ground Improvement Techniques, Bangkok, 1982
AA Balkema, Rotterdam, pp 145–152

Schlosser, F and Juran, I (1981)
"Design parameters for artificially improved soils"
In: *Design parameters in geotechnical engineering*
Proc 7th Eur Conf Soil Mech and Found Engg, Brighton, 1979
British Geotech Soc, London, Vol 5, pp 197–226

Snowdon, R A, Darley, P and Barratt, D A (1986)
An anchored earth retaining wall on the Otley bypass: construction and early performance
TRRL Research Report 62, Transport and Road Research Laboratory, Crowthorne

Temporal, J, Craig, A H, Harris, D H and Brady, K C (1989)
"The use of locally available fills for reinforced and anchored earth"
Proc 13th Int Conf on Soil Mech and Found Engg, Rio de Janeiro, Vol 2, pp 1315–1320

Wynne, C P (1988)
Review of bearing pile types
Piling Guide 1, 2nd edn, CIRIA, London

Yamanouchi, T (1986)
"The use of natural and synthetic geotextiles in Japan"
IEM-ISSMFE Joint Symp on Geotechnical Problems, Kuala Lumpur, pp 82–89

6.2 SOIL NAILING

6.2.1 Definition

Soil nailing is the insertion of steel or glass fibre rods into the face of an excavation or an existing slope to reinforce it. The faces of steep slopes are usually protected by shotcrete and mesh for temporary works and by either cast-in-place concrete or prefabricated panels for permanent applications. For shallower reinforced slopes, the surface can be protected by mesh or geogrid-type reinforcement. Slopes as shallow as 40° have been constructed with soil nails (Everton and Gellatley, 1998).

Where an existing slope or face is being strengthened by reinforcing it with inclusions (which may be termed nails) at an orientation essentially perpendicular to the anticipated shear surface, and which will be subject to bending and shear stresses, these nails may more properly be called soil dowels.

6.2.2 Principle

The rods are passive reinforcement which, when the excavated face begins to strain, generate tensile forces and shearing resistance to counter the yield of the soil.

6.2.3 Description

Soil nailing has been developed in France, Germany and the United States over the past 25 years or so. Considerable worldwide interest in the technique has led to publication of several reviews. These include Guilloux and Schlosser (1985), Nicholson (1986), Bruce and Jewell (1986, 1987a and b, and Jewell 1990), Munfakh *et al*, (1987), Juran and Elias (1987), Gnilsen (1988) and *Recommandations Clouterre 1991* (FHWA, 1993). The nails are generally rods of steel or glass fibre. They are inserted by driving, by drilling and grouting and by pneumatic firing (Myles and Bridle, 1991). Figure 6.11 illustrates the gravity-wall concept that can be applied to soil nailing, where the reinforced soil forms a monolithic block of composite ground. Stocker *et al*, (1979) suggest that this can support its own weight and resist external forces.

Figure 6.12 shows the construction procedure, and typical head details are given in Figure 6.13. Where temporary shotcrete facing is not adopted, corrosion protection at the nail head may be provided by precast or cast-in-place concrete head details. If no continuous facing element is provided (eg shotcrete and mesh) the adequacy of the face plate should be checked against bearing failure. HA 68/94 gives guidance on this check. Bearing tests can be carried out on the excavated face during construction to determine bearing capacity more accurately. Facing construction can consist of shotcrete about 100–250 mm thick, but installed in layers about 80–120 mm thick. The shotcrete is generally reinforced with welded wire mesh. Alternatively, cast-in-place concrete or prefabricated concrete or steel panels can be used for permanent structures. For shallower slopes, the short-term stability of low-height benches in certain soils and rocks can be exploited to avoid the need for shotcreting or mesh. Care has to be taken with this, so pre-construction trials may be necessary to validate the assumptions underlying this approach. Slopes that are not permanently faced are free to slough or ravel between nails, when it may be necessary to install a surface cover of mesh or geo-mesh held in place by short pins. This consideration can often dictate the vertical and horizontal nail spacings.

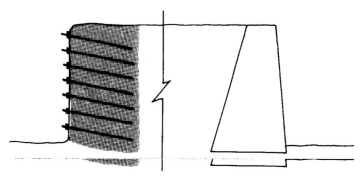

Figure 6.11   *Soil nailing forming a structure like a gravity wall*

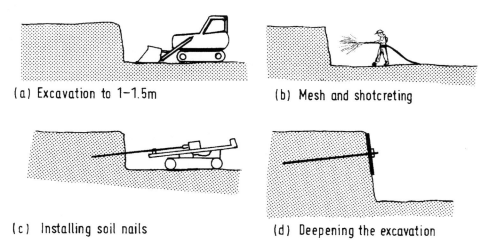

(a) Excavation to 1–1.5m

(b) Mesh and shotcreting

(c) Installing soil nails

(d) Deepening the excavation

Note: operations (b) and (c) are often carried out in reverse order.

Figure 6.12   *Soil nailing construction sequence*

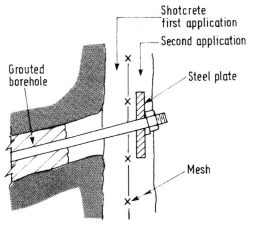

Note: it is common practice to use a single application of shotcrete, with the mesh put against the face. Also not shown are the drains that are an important part of a soil nailing system (see Section 6.2.6).

Figure 6.13   *Soil nail head arrangement*

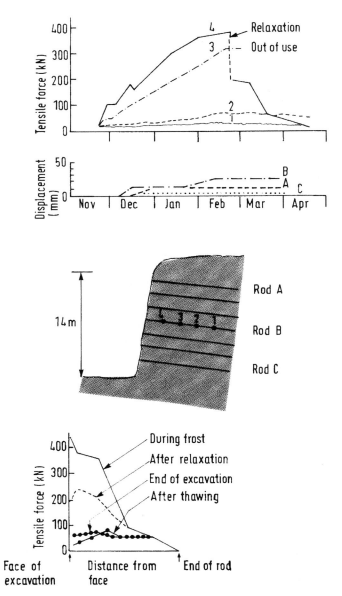

Figure 6.14   *Tensile forces and displacements in an example of soil nailing (after Guilloux et al, 1983)*

CIRIA C573

Although there are similarities between reinforced soil and soil nailing, such as the soil/reinforcement interaction developing as friction along the inclusions, the reinforcement forces being mainly tensile, and the facings for both types of wall being thin, there are three main differences (Schlosser, 1985).

1. Construction procedure. A nailed wall is constructed from the top down, which creates different stress and strain patterns, particularly during construction.

2. Stiffness of reinforcement. Soil nailing reinforcement using a grouted bar can withstand both tensile forces and bending moments.

3. Ground conditions. For a nailed wall, the soil is natural ground (rather than controlled fill) and may be very variable and water bearing.

6.2.4 Applications

Soil nailing has two main applications, both usually being of a temporary nature.

1. Retaining structures. The nails are installed to be almost horizontal and are subjected mainly to tensile forces. Figure 6.14 gives the results of measurements of tensile forces in soil nails and their displacements.

2. Slope stabilisation. Shear and bending stresses develop in the bars. Jewell (1980) has suggested that the greatest effect can be obtained if the reinforcement is inclined at about 60° to the failure surface, corresponding to the direction of maximum tensile strain. This application is shown in Figure 6.15.

American case histories are listed by Bruce (1989) and European ones by Bruce and Jewell (1987 a and b).

Nailing has been used for excavations as deep as 40 m and for stabilising slopes; Munfakh et al, (1987) and Bruce and Jewell (1987 a and b) illustrate other applications, such as ground support in tunnelling.

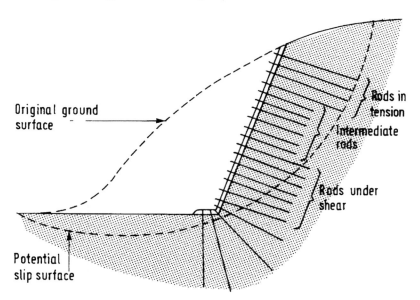

Figure 6.15   *Soil nailing for stabilisation of a cut slope (after Schlosser and Juran, 1981)*

6.2.5 Limitations

Metal soil nails will corrode. BS 8006:1995 requires that the degree of corrosion to be expected over the design life should be assessed and allowed for – this code giving guidance on sacrificial thicknesses for galvanised and non-galvanised nails. There is debate in the industry as to whether soil nails should be doubly protected as is required for ground anchorages (BSI, 1991). The underlying concern is that corrosion of steel bars under shear and bending may not occur evenly (the implicit assumption of providing a sacrificial thickness), but as local, deep pitting. Surrounding the steel bar with grout provides a degree of resistance for drilled nails. The main difference between anchored and nailed support solutions is the number of elements. Accordingly, forces in the individual nails are much smaller than in individual anchors, so the whole nailed assemblage is less sensitive to the failure of an individual member than an anchored structure would be.

As horizontal walings appear not to be used, in near-vertical nailed walls there is little or no ability to transfer lateral pressure should a group of nails fail. The nailed wall has no redundancy. Overall stability has to be checked thoroughly; indeed, slip failures of whole nailed walls have occurred. Two failed and subsequently repaired nailed walls are reported by Bruce and Jewell (1987a). Large creep movements are also likely with soil nails supporting saturated clayey soils (Munfakh et al, 1987).

6.2.6 Design

There is still much discussion about the assessment of the overall stability of a nailed structure. Gnilsen (1988) discusses the differences of approach between Europe and the United States. The design assumptions of Schlosser (1983) and Juran and Beech (1984) are widely used, but see Jewell (1990), Jewell and Pedley (1990 and 1991) and Schlosser (1991) for a discussion of these. Barley (1999) explains the importance of installation effects on nail performance.

Using four fully instrumented retaining structure case histories, Juran and Elias (1987) compared the American and European design methods. Their provisional conclusion was that the behaviour of soil nailed walls was comparable to that of braced open cuts. However, post-construction creep displacements could cause a significant increase in the tension forces in the inclusions.

Bruce and Jewell (1987a and b) examined many case histories using a set of parameters derived to compare different soil conditions, nail types and geometries and relative performance.

It is very important to incorporate vertical drainage not only behind the shotcrete to prevent a build-up of water that would otherwise load the facing, but also deeper drains extending from the front of the wall to the back of the nailed zone.

Continuous bonding distinguishes nails from anchorages. The effect is to reduce nail forces at the face, allowing the use of only a thin cover with the primary function of resisting erosion of the face. Another important effect of continuous bonding along the nail is that the location of the critical failure surface is forced deeper into the slope so that rarely is a Rankine wedge the most critical. This is a source of benefit relative to anchorages.

6.2.7 Controls

Depending on the choice of nailing method, the size, spacing, length and inclination of the nails are primary controls. In that the intention is to achieve a stable cut face, monitoring of any lateral outward movement of the face may be desirable. Construction control would include checking the quality of the grout (if used) and shotcrete and monitoring the drilling techniques. It is also usual to test preliminary nails (which Barley (1999) calls sacrificial production nails), provided they are installed with identical drilling, flushing and grouting methods. Other controls include logging of arisings and, where appropriate, monitoring groundwater conditions to confirm design assumptions.

A valuable safeguard with a soil-nailing system is that additional nails can easily be installed during construction if slope movement occurs or is greater than expected.

6.2.8 References

Barley, A D (1999)
"Controlling factor"
Ground Engineering, Vol 32, No 2, February, pp 12–13

BSI (1989)
BS 8081: 1989 *Code of practice for ground anchorages*
British Standards Institution, London

Bruce, D A (1989)
"American developments in the use of small diameter inserts as piles and in situ reinforcement"
Proc Int Conf on Piling and Deep Foundations, London
Deep Foundations Institute, Vol 1, pp 11–22

Bruce, D A and Jewell, R A (1986) and (1987a)
"Soil nailing: application and practices, Parts 1 and 2"
Ground Engineering, Vol 19, No 8, November, pp 10–15 (Pt 1) and Vol 20, No 1, January, pp 21–38 (Pt 2)

Bruce, D A and Jewell, R A (1987b)
"Soil nailing: the second decade"
Proc Int Conf on Foundations and Tunnels, London, Vol 2, pp 68–83

Everton, S J and Gellatley, G M (1998)
"Innovation and cost saving through design development on the M6 DBFO"
In: *The value of geotechnics*, Proc AGS Seminar at Instn Civ Engrs, London, 4 November 1998, pp 323–332

Federal Highway Administration (1993)
Recommendations Clouterre 1991 for designing, calculating, constructing and inspecting earth support systems using soil nailing
FHWA-SA-93-026, July, US Department of Transportation (English Translation of French National Research Project *Recommandations Clouterre 1991*, Presses Ponts et Chaussées, Paris)

Gnilsen, R (1988)
"Soil nailing debate"
Civil Engineering, Am Soc Civ Engrs, Vol 58, No 8, August, pp 61–64

6.3.3 Description

Figure 6.16 shows the construction procedure for a micro-pile in the UK. Drilling, at different inclinations, is by a rotary method using a heavy-duty outer casing in short lengths fitted with an annular cutting head. This reduces noise and vibration. Either water or bentonite mud is the flushing medium. Depths of 30 m or more can be reached. When the bore is completed it is flushed clean and then filled with grout placed by tremie. Grout is neat cement or sand-cement, and may be injected under pressure. Reinforcement is installed and the temporary casing withdrawn maintaining grout levels as necessary. The reinforcement may be as reinforcing cages, single or groups of high-strength bars, or steel pipes (Bruce, 1989). Micro-piles may also be prestressed to reduce the number required (Mascardi, 1985). Equipment is compact and mobile, can be used in confined spaces or where access is difficult, and is capable of drilling through masonry, concrete, steel, and bouldery ground. At the connection to any structure, adequate bond length must be provided so that loads can be transferred properly. Where micro-piles are acting in tension, or in aggressive ground conditions, corrosive protection may be based on the recommendations of BS 8081:1989.

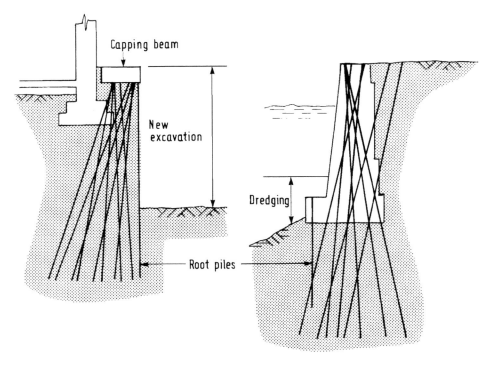

(a)Retaining wall to support existing building   (b) Strengthened foundation

Figure 6.1   *Examples of reticulated root pile structures (after* Ground Engineering, *1983)*

6.3.4 Applications

Figures 6.17 (a) and (b) show how micro-piles can be combined with the ground to form a retaining structure or a strengthened foundation. Note the capping beam connecting the heads of the wall piles and in Figure 6.17 (b) that the existing wall is drilled through to form the piles.

Micro-piles have been used to stabilise slopes in various ground conditions (Figure 6.18). Ellis (1985) describes further examples in the UK of underpinning, forming retaining walls and stabilising slopes; these include a bridge abutment (Figure 6.19) which was underpinned by 170 root piles of 133 mm diameter (4 per metre run). In addition, a reticulated *pali radice* structure was installed in front of the abutment to resist earth pressures and stabilise any potential slip surface: this involved 10 root piles per metre run. Further case histories are described by Dash (1984) and Attwood (1987), including the combination of soil nails and micro-piles to form a reinforced soil wall.

Micro-piles can be used in conjunction with other methods of ground improvement: Miki and Kodama (1985) give an example of root piles below a *Reinforced Earth* wall.

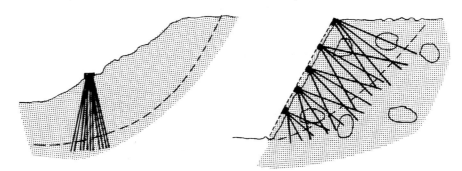

Figure 6.18   *Reticulated root piles for slope stabilisation (after Lizzi, 1985)*

6.3.5   Limitations

Although micro-piles can be installed very close to existing structures, and be drilled through any type of ground, including boulders, the system has two main limitations (Bruce, 1989). First, the costs of a reticulated pile wall can be high and, second, there is as yet no generally agreed design method.

6.3.6   Design

Mascardi (1985) discusses the determination of the ultimate capacity and settlement behaviour of single micro-piles. Proposals for the design of reticulated pile walls were made by Lizzi (1984) and extended by Brandl (1984) using the analogy of reinforced concrete. The main assumption is that the pile wall acts basically as a monolithic mass, with the micro-piles themselves acting as lines of force (Bruce, 1989). For stabilisation of slopes, the micro-piles provide added shearing resistance at a potential failure surface (Bruce, 1989).

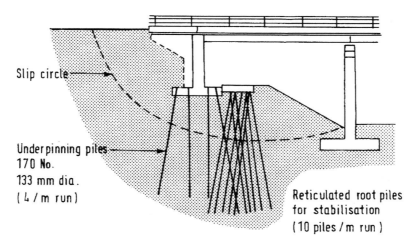

Figure 6.19   *Root piles for underpinning and stabilising an existing abutment (after Ellis, 1985)*

Cantoni *et al* (1989) have proposed an alternative approach that allows the spacing and number of rows of micro-piles required to be determined. The forces on the walls are calculated using the proposals of Ito and Matsui (1975). These proposals are discussed further in Section 6.4 (Slope dowels). Cantoni *et al* (1989) used reticulated micro-pile walls in conjunction with permanent ground anchorages, and give worked examples in their paper. In addition, it is prudent to provide an upper reinforcement cage at the connection with a structure to resist buckling, bursting and shear stresses. There appears to be less wall movement when a capping beam is constructed first in which are cast tubes for the piles.

6.3.7 Controls

Controls are essentially those of piling, but with the special requirement of accuracy in inclined drilling. Checking that the micro-piles have achieved what was intended could involve surveying, eg to monitor levels or face movement, or inclinometry to monitor lateral movements.

6.3.8 References

Attwood, S (1987)
"Pali radice: their uses in stabilising existing retaining walls and creating cast-*in-situ* retaining structures"
Ground Engineering, Vol 20, No 7, October, pp 23–27

Brandl, H (1984)
"Improvement of cohesionless soils"
Proc 8th Eur Conf Soil Mech and Found Engg, Helsinki, 1983
AA Balkema, Rotterdam, Vol 3, pp 1009–1026

Bruce, D A (1989)
"American developments in the use of small diameter inserts as piles and in situ reinforcement"
Proc Int Conf on Piling and Deep Foundations, London
Deep Foundations Institute, Vol 1, pp 11–22

BSI (1989)
BS 8081: 1989 *Code of practice for ground anchorages*
British Standards Institution, London

Cantoni, R, Collotta, T, Ghionna, U N and Moretti, P C (1989)
"A design method for reticulated micro pile structures in sliding slopes"
Ground Engineering, Vol 22, No 4, May, pp 41–47

Dash, V (1984)
"New methods of slope stabilisation"
Proc Int Conf on in situ soil and rock reinforcement, Paris, pp 249-55

Ellis, I W (1985)
"The use of reticulated pali radice structures to solve slope stability problems"
Proc Symp on Failures in Earthworks, Thomas Telford, London, pp 432–435

Ground Engineering (1983)
"Fondedile Foundations comes of age"
Ground Engineering, Vol 16, No 5, pp 23–28

Ito, T and Matsui, T (1975)
"Methods to stabilise lateral force acting on stabilizing piles"
Soils and Foundations, Vol 15, No 4, pp 43–59

Lizzi, F (1984)
"The 'reticulo di pali radice' (reticulated root piles) for the improvement of soil resistance"
Proc 8th Eur Conf Soil Mech and Found Engg, Helsinki, 1983
AA Balkema, Rotterdam, Vol 2, pp 521–530

Lizzi, F (1985)
"The 'pali radice' (root piles): a state-of-the-art report"
Proc Int Symp Recent developments in ground improvement techniques, Bangkok, 1982
AA Balkema, Rotterdam, pp 417–432

Mascardi, C A (1985)
"Design criteria and performance of micro-piles"
Proc Int Symp Recent developments in ground improvement techniques, Bangkok, 1982
AA Balkema, Rotterdam, pp 439–450

Miki, G and Kodama, H (1985)
"Practical uses of the root pile method in Japan"
Proc Int Symp Recent developments in ground improvement techniques, Bangkok, 1982
AA Balkema, Rotterdam, pp 433–438

6.4 SLOPE DOWELS

6.4.1 Definition

Slope dowels are piles or piers that are installed as rigid reinforcement to stabilise an unstable slope. Usually vertical or near vertical, the dowels are separated by substantial gaps, ie they do not form walls. They may be installed in single or several rows. Note that at a smaller scale soil nails (Section 6.2.1) may be installed as dowels.

6.4.2 Principle

Slope dowels are mainly used for slopes which are creeping, a condition defined as moving continuously, but not quite at the point of failure or limit equilibrium. Creep rates can vary between perhaps 0.1 mm/month and 50 mm/month (Gudehus and Schwarz, 1985). The function of the dowels is to reduce the distortion rate in the sheared zone within the slope by providing additional shear and moment capacity (Schlosser *et al*, 1984). Figure 6.20 illustrates this. Dowel spacings are usually two to three times the pile diameter or pier width (Nethero, 1982). It is generally assumed that soil arching takes place between the dowels, in effect forming a wall.

Cantilever or anchored pile walls can be used instead of slope dowels, generally where a pre-existing failure surface means that the slope is completely unstable.

6.4.3 Description

Slope dowels are formed by the conventional techniques of pile and pier construction, as described in Tomlinson (1987) and CIRIA Piling Guide 1 (Wynne, 1988).

6.4.4 Applications

The use of slope dowels to stabilise landslides developed independently in both Europe and Japan after the Second World War. The experiences were brought together in Speciality Session 10 of the Tokyo Conference on Soil Mechanics and Foundation Engineering (1977). Japanese experience was described by Fukuoka (1977), who noted that timber piles had been used to stabilise slopes more than 100 years earlier. Fukuoka made the point that piles should be installed ideally after movement of the slope had been slowed down by drainage, stabilising berms and perhaps with soil removal.

European experience was represented by De Beer (1977) and Sommer (1977), the latter describing the situation shown on Figure 6.21. A 12 m-high embankment was raised on a slope of stiff, plastic clay whose residual angle of shearing resistance had been reduced by progressive sliding to about 8°. The embankment's weight increased the rate of movement. Figure 6.22 shows the results of measurements of earth pressures on the dowels. The measured earth pressures were about 30 per cent of those calculated using Brinch Hansen's (1961) method, and on the upslope side were near to active earth pressures.

A case history described by Cartier and Gigan (1983) is of a landslide stabilised using tubular steel piles, in three rows at the toe of the slope. Winter, Schwarz and Gudehus (1984) give details of three case histories and typical calculations; they also reassess the case history of Sommer (1977) shown in Figures 6.21 and 6.22. Cartier *et al* (1984) give information about three sites where the dowels were formed either by steel sheet piles or by bored piles. They present measurements of movements in another case, where two rows of steel H-piles stabilised a railway embankment above a new motorway. The back analysis of this is described by Guilloux *et al* (1984). Gudehus and Schwarz (1985) presented three more case histories in relation to their proposed method for the design of dowels. Snedker (1985) gives details of one of the few published case histories in the UK.

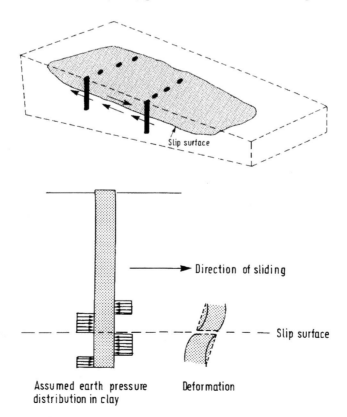

Figure 6.20   *Principles of slope dowels (after Gudehus and Schwarz, 1985, and Sommer, 1977)*

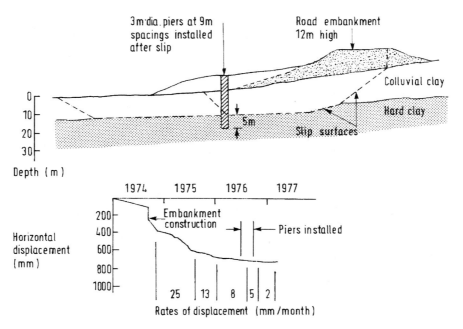

Figure 6.21 *Stabilisation by slope dowels (after Sommer, 1977)*

6.4.5   Limitations

The success of the use of slope dowels depends on the density and distribution of the dowels – soil should not flow between them. Usually, a combination of other measures will be needed as well as the slope dowels. These may include additional drainage, berms and soil removal.

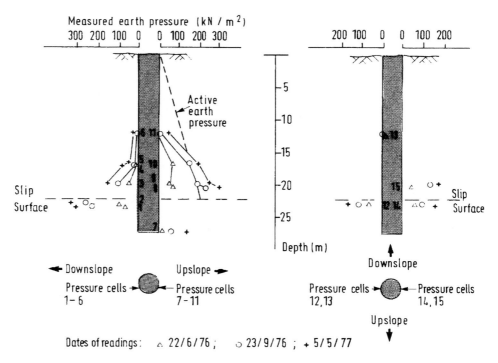

Figure 6.22 *Measurements of earth pressures on a slope dowel (after Sommer, 1977)*

There are several methods of design, discussed briefly below, which are based on very different assumptions and theories of soil behaviour. Comparison of calculations by the various methods is essential to determine the approximate spacing and number of dowels, as there is no widely accepted design procedure as yet.

6.4.6 Design

The early European approach to design was described by De Beer (1977) and by Sommer (1977): it is based on the proposals of Brinch Hansen (1961). Winter *et al* (1984) proposed a method based on a viscosity law for cohesive soils. Design curves are included in the paper. This method was further developed by Gudehus and Schwarz (1985), who calculated curves for deflection, lateral load, transverse force and bending movement for different relative sizes of dowel. Optimum dowel diameters are suggested to be about 5 per cent of the depth to the slip surface.

Yet again, Ito and Matsui (1975 and 1977) and Ito *et al* (1981) suggest a theory based on plastic deformation that does not invoke arching between dowels. Ito *et al* (1981) also address the contribution made by the dowels to overall slope stability. Alternatively, Wang and Yen (1974) proposed that the forces on the dowels comprised the earth pressure at rest upslope of the pile and soil arching pressures. Gray (1978) illustrates the elements of these two theories and compares them. Almost all the various design methods are discussed by Viggiani (1981), who proposed a limit equilibrium method based on the concepts of Broms (1964).

6.4.7 Controls

Construction controls are essentially those required for pile and pier construction. Additional monitoring, eg by surveying and inclinometry, may be required during the course of the work to check slope stability and, later, of the completed dowel system to check that it has achieved its purpose.

6.4.8 References

Brinch Hansen, J (1961)
The ultimate resistance of rigid piles against transversal forces
Bulletin 12, Danish Geotechnical Institute, Copenhagen, pp 5–9

Broms, B B (1964)
"The lateral resistance of piles in cohesionless soils"
J Soil Mech and Found Div, Am Soc Civ Engrs, Vol 90, SM3, pp 123–156

Cartier, G and Gigan, J P (1984)
"Experiments and observations on soil nailing structures"
Proc 8th Eur Conf Soil Mech and Found Engg, Helsinki, 1983
AA Balkema, Rotterdam, pp 473–476

Cartier, G, Delmas, P, Gastin, F and Morbois, A (1984)
"Reinforcement de talus de remblais instables per clouage"
Proc Int Conf on In situ soil and rock reinforcement, Paris, pp 237–242

De Beer, E (1977)
"Piles subjected to static lateral loads"
Proc 9th Int Conf Soil Mech and Found Engg, Tokyo, pp 1–14

Fukuoka, M (1977)
"The effects of horizontal loads on piles due to landslides"
Proc 9th Int Conf Soil Mech and Found Engg, Tokyo, pp 27–42

Gray, D J (1978)
"Role of woody vegetation in reinforcing soils and stabilising slopes"
Proc Symp on Soil reinforcing and stabilising techniques, Sydney, pp 253–306

Gudehus, G. and Schwarz, W. (1985)
"Stabilisation of creeping slopes of dowels"
Proc 11th Int Conf Soil Mech and Found Engg, San Francisco, Vol 3, pp 1697–1700

Guilloux, A, Notte, G and Gouin, H (1984)
"Experiences on retaining structure by nailing in moraine soils"
Proc 8th Eur Conf Soil Mech and Found Engg, Helsinki, 1983
AA Balkema, Rotterdam, pp 499–502

Ito, T and Matsui, T (1975)
"Methods to stabilize lateral force acting on stabilizing piles"
Soils and Foundations, Vol 15, No 4, pp 43–59

Ito, T and Matsui, T (1977)
"The effects of piles in a row on the slope stability"
Proc 9th Int Conf on Soil Mech and Found Engg, Tokyo, pp 81–86

Ito, T, Matsui, T and Hong, W P (1981)
"Design methods for stabilising piles against landslide – one row of piles"
Soils and Foundations, Vol 21, No 1, March, pp 21–37

Nethero, M F (1982)
"Slide control by drilled pier walls"
In: *Application of walls to landslide control problems* (R B Reeves, ed)
Am Soc Civ Engrs, pp 61–76

Schlosser, F, Jacobsen, H M and Juran, I (1984)
"Soil reinforcement"
Proc 8th Eur Conf Soil Mech and Found Engg, Helsinki
AA Balkema, Rotterdam, Vol 3, pp 1159–1180

Snedker, E A (1985)
"The stabilisation of a landslipped area to incorporate a highway by use of a system of bored piles"
Proc Symp on Failures in Earthworks
Thomas Telford, London, pp 436–438

Sommer, H (1977)
"Creeping slope in a stiff clay"
Proc 9th Int Conf Soil Mech and Found Engg, Tokyo, pp 113–118

Tomlinson, M J (1987)
Pile design and construction practice
Viewpoint Publications

Viggiani, C (1981)
"Ultimate lateral load on piles used to stabilise landslides"
Proc 10th Int Conf Soil Mech and Found Engg, Stockholm
AA Balkema, Rotterdam, Vol 3, pp 555–560

Wang, W L and Yen, B C (1974)
"Soil arching in slopes"
J Geotech Engg, Vol 100, No GTI, January, pp 61–78

Winter, H, Schwarz, W and Gudehus, G (1984)
"Stabilisation of clay slopes by piles"
Proc 8th Eur Conf Soil Mech and Found Engg, Helsinki, 1983
AA Balkema, Rotterdam, Vol 2, pp 545–550

Wynne, C P (1988)
Review of bearing pile types
Piling Guide 1 (2nd edn), CIRIA, London

6.5 EMBANKMENT PILES

6.5.1 Definition

Embankment piles are vertical and raking bearing piles installed below embankments to provide additional support when the ground alone cannot carry the loading without excessive settlement or even bearing capacity failure. They are also called relief piles.

6.5.2 Principle

The principle of embankment piles is exactly the same as that of bearing piles: they transfer the superimposed load to a greater depth. As such they are either relief piles to improve stability or settlement-reducing piles. Figure 6.23 illustrates the principle.

6.5.3 Description

Piling for embankments was developed in Sweden in the 1920s and was often used in conjunction with lightweight fill (Broms, 1979). It is an alternative to pre-loading and deep drains (Aldrich, 1964).

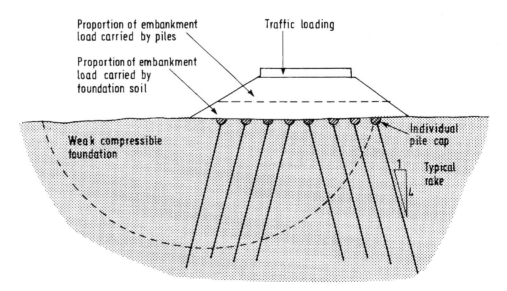

Figure 6.23 *Principle of embankment piles*

Embankment piles can be used to support lengths of embankment or in the transition zone approaching a structure such as a bridge abutment. Figure 6.24 shows how pile depths are altered to provide the transition from embankment to piled bridge. Individual pile caps are often used and cover about 25 per cent of the area supported. In recent years, geotextiles are often laid over the formation to provide tension reinforcement and to assist the arching action between individual pile caps.

6.5.4 Applications

Embankment piles have been used for many years in Scandinavia and in the Far East; there are also several applications in the British Isles. Perhaps the first use here was that reported by Reid and Buchanan (1984) and shown in Figure 6.24. Published examples are those described by Holtz and Massarsch (1976), Vavasour and Ratanaprakarn (1983), Collingwood and Fenwick (1985), Finborud (1985), Chin (1986), and Springett and Stephenson (1988), who give details of a piled raft.

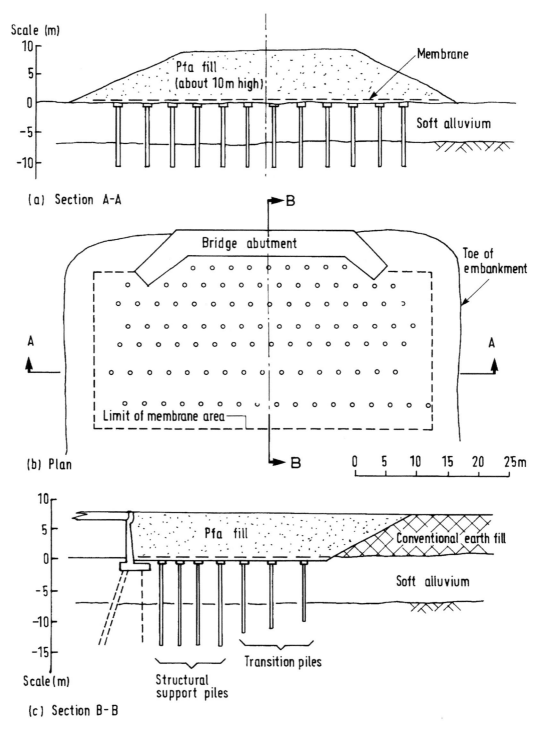

Figure 6.24   *Embankment piles approaching an abutment (after Reid and Buchanan, 1984)*

6.5.5 Limitations

Driving large groups of piles in soft clays increases pore-water pressures and displaces the ground horizontally and vertically upwards. Compressibility of the soft clay can be increased. Consolidation follows, and allowance may have to be made for the additional loads from negative skin friction. The high lateral earth pressures created during pile installation can, in turn, cause the toe of the bank to fail. Pile behaviour should also be ductile. If it is not, and a pile fails, the fill may settle locally because load is not transferred to adjacent piles.

6.5.6 Design

Broms (1979) and Holmberg (1979) discuss the design of embankment piles in some detail. These design methods were supported by a series of model tests (Hewlett and Randolph, 1988), which suggested that the fill acts as a series of domes forming vaults spanning between pile caps. More recently Russell and Pierpoint (1997) made an assessment of design methods for piled embankments. They noted that the methods predict different pile loads and proportions of load taken by reinforcement and that not all current methods are suitable for all embankment geometries. The design of reinforcement to span the pilecaps and transfer the load from the embankment into the piles is covered in BS8006:1991.

6.5.7 Controls

Construction controls are essentially those for piling. The effectiveness of the system could be checked by settlement surveying.

6.5.8 References

Aldrich, H P (1964)
"Precompression for support on shallow foundations"
In: *Design of foundations for control of settlement*, Am Soc Civ Engrs, pp 471–486

BSI (1995)
BS 8006: 1991 *Code of practice for strengthened/reinforced soils and other fills*
British Standards Institution, London

Broms, B B (1979)
"Problems and solutions to construction in soft clays"
Proc 6th Reg Conf Soil Mech and Found Engg, Singapore, Vol 2, pp 27–36

Chin, F K (1986)
"The design and construction of high embankments in soft clay"
Proc 8th South East Asian Geotech Conf, Kuala Lumpur, Vol 2, pp 42–59

Collingwood, R W and Fenwick, T W (1985)
"Selby diversion of the East Coast Main Line: Construction"
Proc Instn Civ Engrs, Vol 77, February, PE I, pp 49–84

Finborud, B (1985)
"Embankment settlement and improvement at the North Luzon Expressway, Philippines"
Proc 8th South East Asian Geotech Conf, Kuala Lumpur, Vol 2, pp 9–15

Hewlett, W J and Randolph, M F (1988)
"Analysis of piled embankments"
Ground Engineering, Vol 22, No 3, April, pp 12–18

Holmberg, S (1979)
"Bridge approaches on soft clay supported by embankment piles"
Geotechnical Engineering, Vol 10, No 1, June, pp 77–89

Holtz, R D and Massarsch, K R (1976)
"Improvement of the stability of an embankment by piling and reinforced earth"
Proc 6th Eur Conf on Soil Mech and Found Engg, Vienna, Vol 1.2, pp 473–478

Reid, W M and Buchanan, N W (1984)
"Bridge approach support piling"
In: *Piling and ground treatment*, Thomas Telford, London, pp 267–274

Russell, D and Pierpoint, N (1997)
"An assessment of design methods for piled embankments"
Ground Engineering, Vol 30, No 10, November, pp 39–44

Springett, M and Stephenson, T D (1988)
"A55 Llandulas to Glan Conwy: design"
Proc Instn Civ Engrs, Vol 84, October, PE 1, pp 939–964

Vavasour, P and Ratanaprakarn, R (1983)
"The design of the Din Daeng-Port section of the first stage Expressway in Bangkok"
Proc Instn Civ Engrs, Vol 74, May, Part 1, pp 225–244

7   Improvement by structural fill

In his1977 Rankine lecture, de Mello expressed one of the approaches to engineering design as "if you do not like your universe, change it!". One of the oldest, but still viable, examples of this philosophy is to replace a weak soil with a better one. Another, more recent option is to use lightweight materials instead of heavier earthfills above weak ground.

These options are described in this section under the following headings:

- removal and replacement (Section 7.1)
- displacement (Section 7.2)
- reducing load (Section 7.3).

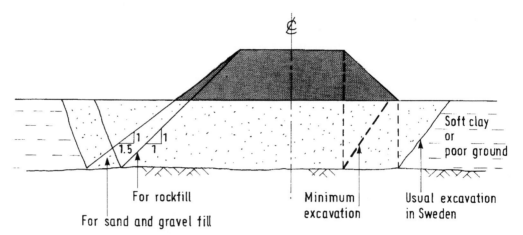

Figure 7.1   *Removal of poor ground and replacement (after Broms, 1979)*

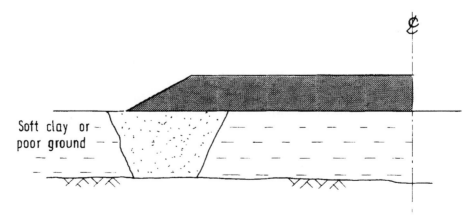

Figure 7.2   *Replacement of poor ground at edges of wide embankments (after Broms, 1979)*

7.1 REMOVAL AND REPLACEMENT

7.1.1 Definition

Poor ground is excavated and removed from the site to be replaced by better-quality compacted fill.

7.1.2 Principle

The fill to replace the excavated material is chosen to have properties desirable for the construction project and its placement is controlled so that these properties are consistently achieved. Usually, the purpose is to provide a stronger, less compressible foundation.

7.1.3 Description

Where soft or filled ground is less than about 6 m deep, it may be economical to excavate it and place sand, gravel or rock fill instead (Broms, 1979). Below the water table it may be necessary to use draglines or dredgers to make the excavation, and the replacement fill would have to be a granular material. In dry excavations, conventional backacters or scrapers can be used, with the replacement fill compacted in layers by vibrating rollers. For granular fills placed through water the vibratory, deep compaction methods (Sections 5.2 and 5.4) would be appropriate.

Figure 7.1 shows what might be taken as the minimum excavation width below an embankment. Particular care has to be taken that excavated side slopes do not collapse, trapping soft material below or within the new fill. Sometimes a minimum excavation line on a drawing is a "payment" line. For very soft soils, excavation widths are increased. In soft peat with a high water table, for example, ground disturbance can extend for tens of metres from the excavation and additional support measures may be necessary.

Where the embankment is wide or the soft soil is relatively thin, it may be sufficient only to excavate along the embankment edges (Figure 7.2). This reduces lateral displacement under the embankment loading, but it will not minimise settlements as full-width replacement does. It may act, however, as a long drain, accelerating consolidation if the *in-situ* clay contains sand and silt layers.

Construction over peat mires can be achieved by complete or partial excavation and replacement. These are illustrated in Figures 7.3 (a) and (b) and are described in *Soil Mechanics for Road Engineers* (Road Research Laboratory, 1952) – although the trench-shooting and toe-shooting methods of blasting also there described are unlikely to be used nowadays. It is hard to be sure of complete excavation, especially in thick deposits below the water table. Allowance should be made for some contamination of the replacement fill.

Hydraulic fills may be the chosen replacement material, accepting that after placement they may have to be improved. The definitive paper is probably that by Whitman (1968), who describes case histories of the use of fairly clean sand, silty or clayey sands, silty clays and soft clays (slurry). Eight further case histories, referring generally to fine to medium sands, are reported by Turnbull and Mansur (1973). They advised that improvement can be achieved using the vibratory methods described in Chapter 5. Charles (1987), in a more recent review, agreed, and added pre-loading to the processes of vibration. Cohesive hydraulic fills are the most difficult to improve.

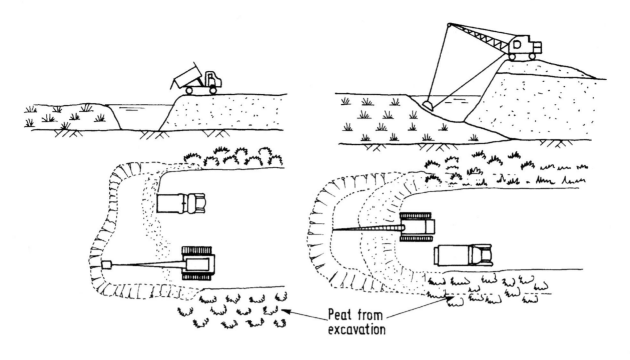

(a) Complete excavation

(b) Partial excavation and partial displacement

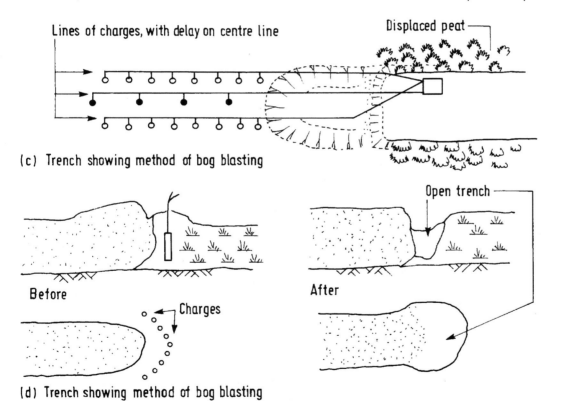

Lines of charges, with delay on centre line

Displaced peat

(c) Trench showing method of bog blasting

Before

Charges

After

Open trench

(d) Trench showing method of bog blasting

Figure 7.3   *Construction over peat mires (after Road Research Laboratory, 1952)*

7.1.4   Applications

Thorburn and MacVicar (1968) describe the use of layers of compacted fill on many sites in the Glasgow area in the 1960s. Excavation was carried out in dry conditions and the sides of the excavation remained stable during compaction. Proximity of adjoining properties limited work in some cases. All unsuitable ground was removed and replaced with compacted granular fill. This was placed in 275 mm layers densified to 225 mm with a 68 kN smooth-wheel vibrating roller. Fill materials included slag, well-burnt colliery shale, industrial boiler ash, pulverised fuel ash, crushed rock and sand.

On a site in Germany, 600 000 m³ of silt was excavated and replaced with gravel compacted by vibratory rollers. A footing test was carried out to assess likely foundation behaviour (Hilmar *et al*, 1984).

Schnabel and Martin (1983) decided to excavate 3–9 m of soft clay overlying karstic limestone. The soft clay was replaced by well-graded aggregate placed in 0.3 m lifts and compacted with a vibrating drum roller. Subsequent column loads bearing on the fill were about 6700 kN for which the settlements were about 12–25 mm after three years.

Gaba and Hyde (1987) describe the excavation of 2.65 m fill and its replacement with compacted granular fill on a site next to the River Thames. A trial embankment established compaction controls and a trial footing load test was also carried out.

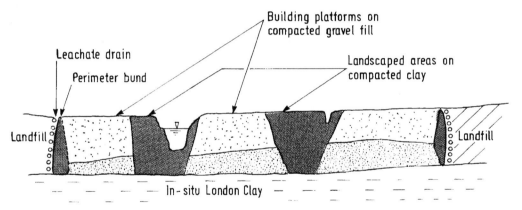

Figure 7.4   *Structural fill replacing landfill (after Gordon et al, 1987)*

In redeveloping a large contaminated landfill site in West London, some 2 700 000 m³ of material were removed and replaced by both granular and clay fills (Gordon *et al*, 1987). The section (to an exaggerated vertical scale) through part of this scheme given as Figure 7.4 shows building platforms formed of gravel fill and between them landscaped areas on clay fills.

A good case study was reported by Everton and Gellatley (1998). This was on the M6 Motorway at Beattock Summit built under a design, build, finance and operate contract. Some 4 m of peat was excavated and replaced with rockfill for a length of 1.5 km close to and without disturbing the West Coast Main Line railway, which was itself constructed on the peat. The work was successfully completed using the Observational Method to monitor the excavation-induced movements and control the new construction operations, without recourse to expensive temporary works such providing sheet piling support to the railway.

7.1.5 Limitations

The disposal of the excavated soil is an important, and perhaps costly, consideration. Very soft clays, completely remoulded or even slurried in the excavation process, may have to be treated specially for disposal (eg by lime stabilisation). The importance of achieving sufficiently stable side slopes of the excavations and the possible extent of ground disturbance are noted in Section 7.1.3 above. In some circumstances, depending on the type of replacement fill, it may have to be densified by vibratory techniques of deep compaction.

7.1.6 Design

Design is essentially a matter of balancing quantities and costs in comparison with methods that leave the poor ground in place. Some guidance is given in Monahan (1986) and in the proceedings of the conference on compaction technology (Thomas Telford, 1987).

7.1.7 Controls

The controls are broadly those that would apply to earthworks operations. Environmental regulations apply to the disposal of the unsuitable material (see Kwan *et al*, 1997, for a discussion of these in the context of the management of ground engineering spoil).

7.1.8 References

Broms, B B (1979)
"Problems and solutions to construction in soft clays"
Proc 6th Reg Conf Soil Mech and Found Engg, Singapore, Vol 2, pp 27–36

Charles, J A (1987)
Review paper. In: *Building on marginal and derelict land*
Thomas Telford, London, pp 95–109

Everton, S J and Gellatley, G M (1998)
"Innovation and cost saving through design development on the M6 DBFO"
In: *The value of geotechnics*, Proc AGS Seminar at Instn Civ Engrs, London,
4 November 1998, pp 323–332

Gaba, A R and Hyde, R B (1987)
"The redevelopment of a former industrial site at Morgan's Walk, Battersea, London"
In: *Building on marginal and derelict land*
Thomas Telford, London, pp 539–558

Gordon, D L, Lord, J A and Twine, D (1987)
"The Stockley Park Project"
In: *Building on marginal and derelict land*
Thomas Telford, London, pp 359–379

Hilmar, K, Knappe, M, Antz, H and Stark, D (1984)
"Ground improvement by soil replacement"
Proc 8th Eur Conf Soil Mech and Found Engg, Helsinki, 1983
AA Balkema, Rotterdam, Vol 1, pp 37–43

Kwan, J C T, Sceal, J S, Bryson, T E, Stanbury, J, Bickerdike, J and Jardine, F M (1997)
Ground engineering spoil; good management practice
Report 179, CIRIA, London

Monahan, E J (1986)
Construction of and on compacted fills
John Wiley and Sons, New York

Road Research Laboratory (1952)
Soil Mechanics for Road Engineers
HMSO, London

Schnabel, J J and Martin, R E (1983)
"Parking garage supported on fill"
J Contn and Engg Mgmt, Am Soc Civ Engrs, Vol 109, No 3, pp 286–296

Thomas Telford (1987)
Compaction Technology, Proc Conf of New Civil Engineer
Thomas Telford, London

Thorburn, S and MacVicar, R S L (1968)
"Soil stabilisation employing surface and depth vibrators"
Structural Engineer, Vol 46, No 10, October

Turnbull, W J and Mansur, C I (1973)
"Compaction of hydraulically placed fills"
J Soil Mech and Found Divn, Am Soc Civ Engrs, Vol 99, No SM11, pp 939–955

Whitman, R V (1968)
"Hydraulic fills to support structural loads"
In: *Placement and improvement of soil to support structures*
Am Soc Civ Engrs, pp 169–193

7.2 DISPLACEMENT

7.2.1 Definition

Soft ground is displaced by a mass of imported granular fill.

7.2.2 Principle

The imported fill is placed in a way that continually exceeds the bearing capacity of the *in-situ* soft soil, which in shearing is remoulded, often to a very low strength, so that a dynamic process of filling, penetration and displacement can be achieved. Mud waves are generated, ahead and to the side of the fill, which tend to act as stabilising berms and may have to be excavated ahead of the fill.

7.2.3 Description

The displacement method has been used in Sweden, both in soft clays and peats, since the 19th century. Broms (1979) suggested that the technique would be economical for thicknesses of soft soil of about 5 m. A maximum depth limit is probably about 15 m. The concepts of displacement methods are shown in Figure 7.5. As there is often a desiccated crust or vegetation strengthening the surface of soft clays this may have to be broken up or stripped. Control of the mud waves is important, partly because of their stabilising effect, but also to avoid soft soil being trapped below the fill, which would subsequently cause uneven settlement. The direction of the mud flow can be adjusted by different leading-edge shapes of the filling (Figure 7.6).

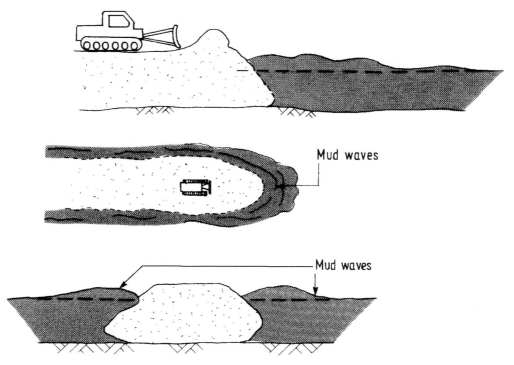

Figure 7.5   *Displacement of soft soils*

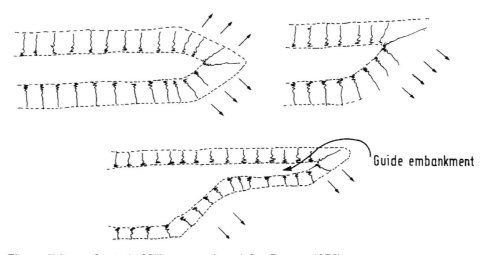

Figure 7.6   *Control of filling operations (after Broms, 1979)*

7.2.4   Applications

The displacement method is probably more used when forming embankments or dikes
across lagoons or bays of lakes or the sea, rather than across land. The technique is used
in Hong Kong and elsewhere to form perimeter bunds for areas of reclamation filling.
Other examples are embankments for roads and railways (eg Webb, 1985).

7.2.5   Limitations

One of the disadvantages of the displacement method is that large volumes of imported
fill are needed. The shape, and hence volume, of fill can be very much larger than the
nominal (minimal) dimensions intended, especially if the underlying firm material has
an irregular or sloping surface or if there are high pore-water pressures in sand layers
within or below the soft clay.

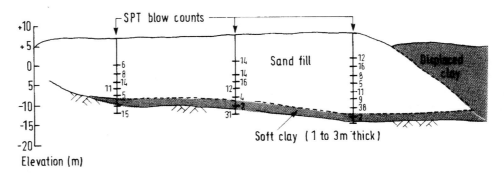

Figure 7.7   *Clay trapped below fill (after Webb, 1985)*

Broms (1979) also points out further disadvantages.

1.   Very large areas can be affected by the displaced soil.

2.   The embankment will settle after construction because of consolidation of soils trapped below or within the fill (see Figure 7.7).

3.   Mud waves can extend to a distance of up to 10 times the thickness of the soft layer. Structures closer than four to five times the thickness may be affected.

7.2.6   Design

There is no generally accepted guidance available. It is usual to develop the filling procedures by trials. As it is usually impracticable to eliminate trapped mud pockets below the fill, the structural design of any building on the fill should be such as to be capable of accommodating a reasonable amount of settlement and differential settlement, as would be realistic in the circumstances.

7.2.7   Controls

The controls usual for earthworks do not normally apply other than care in the selection of fill material. Depending on the site circumstances, it may be necessary or desirable to balance the rate of filling so that it is of sufficient mass and impetus to displace the mud, but at not so violent a rate as to generate untoward mud waves and over-filling.

7.2.8   References

Broms, B B (1979)
"Problems and solutions to construction in soft clays"
Proc 6th Reg Conf Soil Mech and Found Engg, Singapore, Vol 2, pp 27–36

Road Research Laboratory (1952)
Soil Mechanics for Road Engineers
HMSO, London

Webb, D L (1985)
"Construction of a railway embankment by displacement of deep soft clays and silts"
Proc 11th Int Conf Soil Mech and Found Engg, San Francisco, Vol 3, pp 1761–1766

7.3   REDUCING LOAD

7.3.1   Definition

Lightweight fills are used instead of conventional earth or rock fill. The lightweight fills include pulverised fuel ash, expanded shale, clinker, sawdust and, lightest of all, expanded polystyrene blocks.

7.3.2   Principle

Lightweight fills give a greater margin of safety for stability and generate less settlement than conventional fill.

7.3.3   Description

Reducing the applied load can be the simplest and cheapest method of improving stability and reducing settlements (Broms, 1979). Light materials can be expanded shale, slag or sawdust. Finborud (1985) reports the use of lightweight fills weighing about 10 kN/m^3. Table 7.1 lists the relative densities of some lightweight fill materials and some factors affecting their use. The technique was usually limited to minor roads for which the delayed deformations of the lightweight mass would be acceptable. However, the idea was revived in Norway in the 1970s with the use of expanded polystyrene, which has a mass of about 30 kg/m^3.

Table 7.1   *Lightweight fill materials (Delmas et al, 1987)*

Material	Relative density	Comments
Pine or fir bark	0.8–1.0	Compaction difficult; long-term deformation up to 10 per cent; pollutes groundwater
Sawdust	0.8–1.0	Should be kept under water or sealed with asphalt or plastic sheets
Dried, milled and compressed peat	0.2–1.0	Used successfully in Ireland
Pulverised fuel ash	1.0–1.4	Usually above water table; can set; water absorption may increase density
Expanded clay or shale	0.5–1.0	Good physical and mechanical properties
Expanded polystyrene	0.02–0.03	Sensitive to hydrocarbons; good mechanical properties
Extruded polypropylene	0.03	Sensitive to ultra-violet; good mechanical and chemical properties

7.3.4   Applications

Expanded polystyrene blocks forming lightweight fills can be used in the transition zone between an embankment on soft ground and a piled bridge abutment. Differential deformations and lateral pressures on the piles can be limited by the use of polystyrene. It can also be used to enlarge an existing embankment or repair a landslide.

Clarken (1986) and Delmas *et al* (1987) give practical guidance on the construction of embankments of expanded polystyrene.

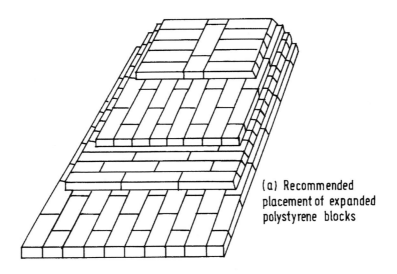

(a) Recommended placement of expanded polystyrene blocks

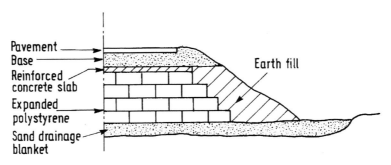

Pavement
Base
Reinforced concrete slab
Expanded polystyrene
Sand drainage blanket
Earth fill

(b) Typical road embankment cross section

Figure 7.8   *Expanded polystyrene for embankment construction*

Figure 7.8a indicates the recommended scheme for placement of the blocks. Figure 7.8b shows a typical cross-section of road embankment using the expanded polystyrene. The first layer of blocks is placed on a sand layer about 500 mm thick.

Lightweight fill was used at Great Yarmouth (Figure 7.9) because the ground could not be pre-loaded for long enough to achieve the settlements required (Williams and Snowdon, 1990; Boyce, 1986). Further Norwegian case histories are reported by Aaböe (1986). Polystyrene has been used beneath the track of tramways in France, and in Japan it has been used to form embankments and to reduce the vibration from trains.

7.3.5   Limitations

Expanded polystyrene must be protected from the hydrocarbons and oil from the road above otherwise it will dissolve and collapse rapidly. Clarken (1986) gives details of the resistance of polystyrene to various chemicals or solvents. Both the upper surface and the sides of the polystyrene mass may have to be protected (Delmas *et al*, 1987).

With some lightweight fills especially the organic ones, precautions may have to taken to prevent damage by burrowing animals.

7.3.6 Design

Where it is a viable engineering option, lightweight fill is likely to be more environmentally acceptable than excavation and replacement as it will cause less disturbance to wetland habitats.

The height of the lightweight fill layer is determined using conventional soil mechanics design methods, allowing for any thickness of normal fill and the road pavement. For polystyrene, wind forces during construction and vehicle impact should be considered. An important factor affecting pavement thickness is that heavy rollers cannot be used for compaction. Thicknesses of 400–600 mm, including 100 mm in reinforced concrete, are required for axle loads of about 100 kN. Bituminous mixes of similar thickness may be required for axle loads of 130 kN or greater.

7.3.7 Controls

It is likely that these have to be developed for each usage and lightweight fill, eg in determining placement and compaction methods and standards.

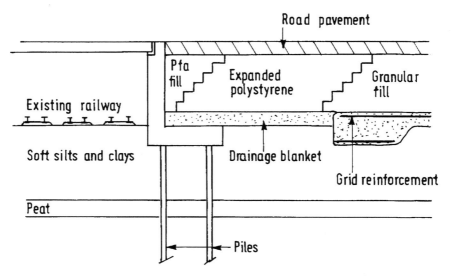

Figure 7.9   *Cross-section of embankment approach to bridge abutment, Great Yarmouth (after* Ground Engineering, *1986)*

7.3.8 References

Aaböe, R (1986)
"Plastic foam in road embankments"
Ground Engineering, Vol 19, No 1, pp 30–31

Boyce, J C (1986)
"Great Yarmouth By-pass – earthworks and pavement design"
J Instn Highways and Transportn, Vol 33, No 1, January, pp 3–8

Broms, B B (1979)
"Problems and solutions to construction in soft clays"
Proc 6th Reg Conf Soil Mech and Found Engg, Singapore, Vol 2, pp 27–36

Clarken, W (1986)
"Expanded polystyrene for road construction"
Proc Sem on lightweight fill in road construction, Dublin, November, pp C1–12

Delmas, P, Magnan, J-P and Soyez, B (1987)
"New techniques for building embankments on soft soils"
Chapter 6 in: *Embankments on soft clays*
Special Publication, Bulletin of the Public Works Research Center, Athens, pp 323–356

Finborud, B (1985)
"Embankment settlement and improvement at the North Luzon Expressway, Philippines"
Proc 8th South East Asian Reg Conf: Soil Mech and Found Engg, Kuala Lumpur
Vol 2, pp 9–15

Williams, D and Snowdon, R A (1990)
A47 Great Yarmouth western bypass: performance during the first three years
TRL Contractor's Report 211, Transport Research Laboratory, Crowthorne

8   Improvement by admixtures

The use of admixtures, such as lime, cement, oils and bitumens, and even sulphur, is one of the oldest and most widespread methods of improving a soil. Usually the purpose is to strengthen a locally available earth fill to construct a low-cost road base, eg cement-stabilised soil or soil-cement, or to mix lime into highly plastic clays. Plant was developed either to mix the stabiliser in place, ie to strengthen subgrades or layers of the fill, or for central mixing to which the soil is transported.

Deep methods of mixing are now available that enable columns of stabilised soil to be formed in the ground. The Swedish technique (lime columns) is described in Section 8.1, lime and cement columns (referred to as the Japanese technique, although it has been available in northern Europe for some 10 years) in Section 8.2, and a deep mix-in-place treatment of ground in Section 8.3. The use of lime for slope stabilisation, introduced in Section 8.4, is an interesting variation of admixture improvement. Sections 8.5 and 8.6 summarise the mix-in-place stabilisation of soil by lime and cement to form a capping layer or a road base.

Lime is produced by burning calcium carbonate rocks (limestones) to form calcium oxide (quicklime), which, when hydrated, is called slaked lime (calcium hydroxide). Lime is used for clay soils, with which it reacts chemically and physically. First, the lime agglomerates the clay particles into friable silt- and sand-size groups through base exchange of cations. Secondly, the lime reacts with the clay minerals to form calcium silicates and aluminates to harden the mixture. It should be noted that the lime used by farmers is pulverised limestone, ie calcium carbonate, and is not suitable for soil stabilisation.

When cement is mixed with soil it hydrates and sets, either bonding soil grains together or forming a skeletal structure of hardened soil around lumps of untreated soil. These two basic soil-cement structures are shown in Figure 8.1.

(a) Cement coating around lumps of clay

(b) Cement bonds at grain contacts in sand and gravel

Figure 8.1   *Structures of soil-cement*

8.1 LIME COLUMNS (SWEDISH METHOD)

8.1.1 Definition

The Swedish method of forming lime columns is the introduction of quicklime by hollow stem auger at depth so that it will be mixed with soft, fine-grained soil as the auger is rotated and raised.

8.1.2 Principle

Soft, clayey soils are strengthened by mixing with lime, possibly being as much as 10 to 50 times stronger after a year. Column-like units of the soil-lime mixture reinforce the ground and may individually be capable of bearing loads of 50–100 kN. In addition to the columns, the lateral pressures in the ground are increased by the volume increase of the column as the lime/soil mixture is introduced. The columns are designed to act not as weak piles but as a composite with the surrounding soil.

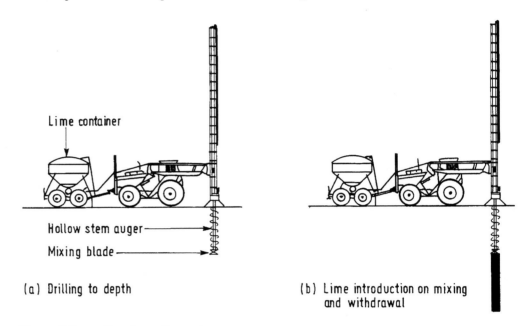

Lime container

Hollow stem auger

Mixing blade

(a) Drilling to depth

(b) Lime introduction on mixing and withdrawal

Figure 8.2 *Forming a lime column*

8.1.3 Description

The lime column process uses quicklime in soft, fine-grained soils to form columns, piers or walls. Parallel development of the technique took place in Sweden – described by Broms and Boman (1979) – and Japan. The method was developed as an alternative to piles as foundations for houses founded on clays with shear strengths of about 10–20 kN/m^2. The columns are about 0.5 m in diameter and up to 15 m deep. A hollow-stem auger with a special blade for mixing is drilled into the ground to the required depth (Figure 8.2). Rotation is then reversed, so that soil is not lifted on the flights of the auger, and the tool is withdrawn slowly at about 25 mm/revolution to ensure thorough mixing of lime and soil. During extraction, the lime is pumped down the hollow stem by compressed air through a hole just above the auger blade to mix with the soil. To prevent clogging of the discharge point, very pure lime is used, with a maximum particle size of less than 0.2 mm. Production is at the rate of up to 1 m/minute, which can produce up to 50 columns/day. Pre-drilling is necessary for rockfills or bouldery soil. Topsoil should also be removed before installation of the lime columns.

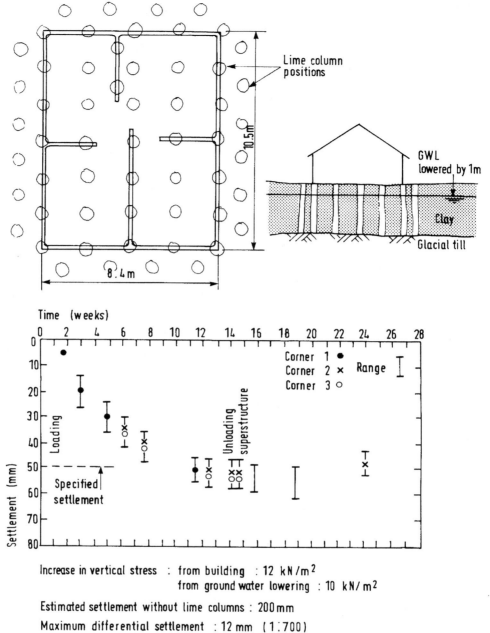

Figure 8.3 *Settlement of a building founded on lime columns (after Holm et al, 1981)*

8.1.4 Applications

Lime columns are appropriate for soils containing at least 20 per cent of clay, and the content of silt and clay should be at least 35 per cent (Broms, 1984). The plasticity index should also be greater than 50 per cent. Gypsum can be added to help stabilise organic soils with moisture contents of up to 180 per cent. According to Holm *et al* (1983a), when gypsum is added to the lime, undrained strength can be three times that when lime is used alone. Lime columns are particularly effective where there is a high ground temperature, because the rate of hardening of the columns is faster. Alkaline (high pH) soils are also advantageous.

Case histories of the Swedish method include that of Holm *et al* (1982), who describe the behaviour of a single-storey house on lime columns over a three-year period. Figure 8.3 shows the column layout and the settlement of the house. The initial undrained shear strength of the soil was about 10 kN/m^2. Holm *et al* (1982) give details of a penetro-meter they developed to assess the strength of the lime columns, and they present results of laboratory and field tests showing the gain in strength from 10 kN/m^2 to more than 200 kN/m^2 in the year following stabilisation. Ahnberg *et al* (1989) compare laboratory and field tests, which suggest that unconfined compression tests and penetrometer tests can lead to conservative design.

A full-scale trial embankment is described by Holm *et al* (1983b) in which a surcharged embankment was constructed on 10 m or so of varved clays. One length was built on lime columns; the settlement was less than half of the settlement of that part of the embankment built over vertical sand drains. The lime columns also apparently compared well with the trial length of embankment piles (Section 7.5).

8.1.5 Limitations

A disadvantage of lime columns is that they can act as drains, and their bearing capacity can decrease with time because of leaching by slightly acidic groundwaters. The lime/clay mixture may be more frost-sensitive than soil on its own. Sometimes the column material appears as matchbox-size lumps, a result of variations in chemical reaction. Alternatively, it can be cracked in layers at every 20–50 mm and be weakest in the centre (Broms, 1979). For these reasons, tests on laboratory mixtures often do not compare well with samples from the field or field tests.

8.1.6 Design

The design of deep-mixed lime columns has been formalised by the publication of a design handbook (Broms, 1984) entitled *Stabilisation of soil with lime columns* and by the Swedish Geotechnical Society Report SGF4:95E *Lime and lime-cement columns.*

8.1.7 Controls

These should relate first to the component materials and their storage, handling and mixing. Testing for rheological characteristics and strength development in the mixed materials should be a matter of course. Careful workmanship involves control over the injected volumes in combination with management of the drilling rates and rotation speeds, etc. and drilling depth. Load-testing the columns once set is impracticable because the load capacity would be very high, but checking the quality of a completed column could be done by both probing and penetration testing while the column is still relatively fresh. Coring would be possible if the project warranted it. Care may be needed to check the overlap of adjacent blocks if they are to form a water barrier.

Broms (1984) recommends the following for construction control of lime columns:

- screwplate or plate loading tests (to determine creep behaviour)
- cone penetration tests to check variations with depth
- use of a sampler of the completed column diameter (eg 0.6 m).

Details of the methods of test and their interpretation are given in Broms (1984). Since then, new and more effective mixing tools, and a penetrometer for testing lime columns have been developed.

8.1.8 References

Ahnberg, H, Bengtsson, P E and Holm, G (1989)
"Prediction of strength of lime columns"
Proc 12th Int Conf Soil Mech and Found Engg, Rio de Janeiro, Vol 2, pp 1327–1330

Broms, B B (1979)
"Problems and solutions to construction in soft clays"
Proc 6th Reg. Conf Soil Mech and Found Engg, Singapore, Vol 2, pp 27–36

Broms, B B (1984)
Stabilisation of soil with lime columns
Design Handbook, 2nd edn, Royal Inst of Technology, Soil and Rock Mechanics Dept, Oslo

Broms, B B and Boman, P (1979)
"Lime columns – a new foundation method"
J Geotech Engg Div, Am Soc Civ Engrs, Vol 105, No GT4, April, pp 539–556

Holm, G, Bredenberg, H and Broms, B B (1981)
"Lime columns as foundation for light structures"
Proc 10th Int Conf Soil Mech and Found Engg, Stockholm, 1981
AA Balkema, Rotterdam, Vol 3, pp 687–694

Holms, G, Trank, R and Ekstrom, A (1983a)
"Comparing lime column strength with gypsum"
Proc 8th Eur Conf Soil Mech and Found Engg, Helsinki, 1983
AA Balkema, Rotterdam, Vol 2, pp 903–907

Holm, G, Trank, R, Ekstrom, A and Torstensson, B A (1983b)
"Lime columns under embankments – a full scale test"
Proc 8th Eur Conf Soil Mech and Found Engg, Helsinki, 1983
AA Balkema, Rotterdam, Vol 2, pp 909–912

Swedish Geotechnical Society (1995)
Lime and lime-cement
Report SGF4:95E, Swedish Geotechnical Society

8.2 LIME AND CEMENT COLUMNS (JAPANESE METHOD)

8.2.1 Definition

The Japanese methods of forming columns in clay soils and sands use quicklime or cement introduced and mixed at depth by drill casing fitted with mixing blades. For offshore work in soft marine clays and sandy soils, a group of, say, eight drill units operate to form blocks of stabilised clay (Figure 8.4).

The technique is called deep soil mixing or deep chemical mixing. Massive stabilised structures can be formed by changing the configuration of the drill units and the positioning of the blocks (Figure 8.5). For onshore work the method is essentially the same as that described in Section 8.1 as the Swedish method.

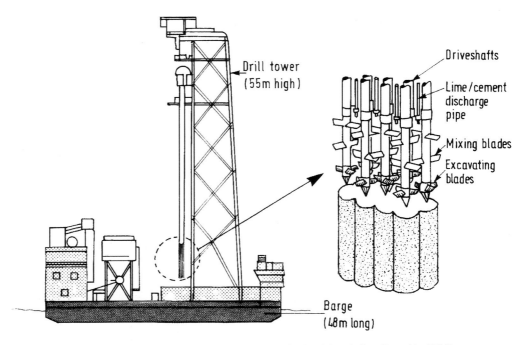

Figure 8.4   *Block formation by deep chemical mixing (after Suzuki, 1985)*

8.2.2   Principle

Soils are strengthened by mixing with lime or cement to form single columns or blocks of stabilised soil.

8.2.3   Description

The Japanese developed the technique of deep chemical mixing with quicklime in order to stabilise deep deposits of soft marine clays offshore (Okumura and Terashi, 1975; Terashi *et al*, 1979). Figure 8.4 gives an idea of the plant on the drill barge and the mixing equipment used to form the lime-stabilised blocks. Kawasaki *et al* (1982) and Suzuki (1985) describe how cement can be used instead of lime with essentially similar equipment. In forming the zones of lime or cement blocks, a 250 mm overlap of mixing is usual. Suzuki (1985) gives typical details of soil conditions, performance of the mixer and case histories with cement mixing. Ordinary Portland or Portland blast-furnace slag cement was used. Table 8.1 gives some typical installation records. Hardening additives are used to assist solidification of the soil/cement mixture. Saitoh *et al* (1985) compared the properties of cement and lime (both slaked and unslaked) concluding that cement was the more effective admixture.

Table 8.1   *Installation of data of deep chemical mixing (after Suzuki, 1985)*

Depth of treatment	to 30 m or more
Plan area of treatment block	4.3–5.7 m^2
Rate of penetration	1–2 m/min
Rotation speed during penetration	20–30 rpm
Rate of withdrawal (mixing stage)	0.5–1.5 m/min
Rotation speed during withdrawal	40–60 rpm

Note: Suzuki (1985) gives examples of eight sites from which these data are drawn. Clays varied from very soft to very stiff and the stabilisation included layers that were mainly of sand.

Japanese onshore use of lime columns is comparable to the Swedish usage, and is described by Kitsugi and Azakami (1985). Again, quicklime is employed in conjunction with powdered additives such as calcium silicate and/or calcium aluminate. These accelerate the pozzolanic reaction between the lime and clay. As the quicklime hydrates with the groundwater, there is a considerable volume increase. The heat of hydration gives rise to high temperatures at the centre of the pile, which have been measured at 300–400°C for up to three hours after mixing. The construction procedure and plant used for single columns are essentially that shown in Figure 8.2.

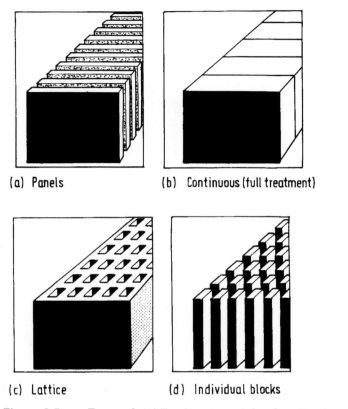

(a) Panels   (b) Continuous (full treatment)

(c) Lattice   (d) Individual blocks

Figure 8.5   *Types of stabilised structure (after Suzuki, 1985)*

8.2.4   Applications

The offshore use by the Japanese of deep chemical mixing is illustrated in Figure 8.6. Here the stabilised foundation supports a cofferdam bounding an area of reclamation.

The main usages of lime columns for onshore purposes in Japan were reported by Kitsugi and Azakami (1985) to be:

- preventing heave in excavations (30 per cent)
- increasing bearing capacity (30 per cent)
- preventing failure of sides of excavations (20 per cent)
- preventing sliding failure of embankments (10 per cent).

8.2.5   Limitations

For onshore use, the limitations of lime columns are likely to be those described in Section 8.1.5 for the Swedish method. The use of cement extends the range to sandy soils, and the offshore system of a bank of drills adds considerably to the possibilities for the technique. The main constraint is that of obstructions such as boulders.

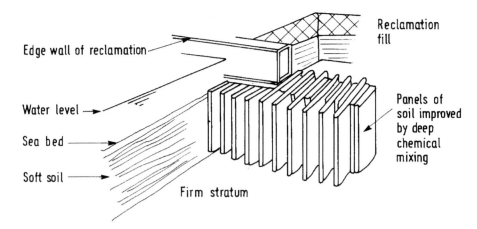

Figure 8.6　*Cutaway section of a stabilised foundation at a reclamation boundary (after Suzuki, 1985)*

8.2.6　Design

The design of offshore applications are described by Terashi *et al* (1979) and Suzuki (1985). Where cement is to be used instead of lime, reference should be made to Kawasaki *et al* (1981) and Saitoh *et al* (1985). For onshore use, guidance is given by Kitsugi and Azakami (1985).

8.2.7　Controls

These should relate first to the component materials and their storage, handling and mixing. Testing for rheological characteristics and strength development in the mixed materials should be a matter of course. Careful workmanship involves control over the injected volumes in combination with management of the drilling rates and rotation speeds etc and drilling depth.

Construction control includes the monitoring of depth, rotation speeds of the mixer blades, penetration and withdrawal speeds, and hydraulic pressure for the excavating and agitating drives. The hoisting load and discharged quantity of the admixture are also monitored. Load-testing the columns once set is impracticable as the load capacity would be very high, but checking the quality of a completed column could be done by a combination of probing or penetration testing while the column is still relatively fresh. Coring would be possible if the project warranted it. Care may be needed to check the overlap of adjacent blocks if they are to form a water barrier.

8.2.8　References

Kawasaki, T, Niina, A, Saitoh, S, Suzuki, Y and Honjyo, Y (1981a)
"Deep mixing method using cement hardening agent"
Proc 10th Int Conf Soil Mech and Found Engg, Stockholm, 1981
AA Balkema, Rotterdam, Vol 3, pp 721–724

Kitsugi, K and Azakami, H (1985)
"Lime column techniques for the improvement of clay ground"
Proc Symp on Recent developments in ground improvement techniques, Bangkok, 1982
AA Balkema, Rotterdam, pp 105–111

Okumura, T and Terashi, M (1975)
"Deep lime mixing method of stabilisation for marine clays"
Proc 5th Asian Reg Conf Soil Mech and Found Engg, India, Vol 1, pp 69–75

Saitoh, S, Suzuki, Y and Shirai, K (1985)
"Hardening of soil improved by deep mixing method"
Proc 11th Int Conf Soil Mech and Found Engg, San Francisco, Vol 3, pp 1745–1748

Suzuki, Y (1985)
"Deep chemical mixing method using cement as hardening agent"
Proc Symp on Recent developments in ground improvement techniques, Bangkok, 1982
AA Balkema, Rotterdam, pp 299–340

Terashi, M, Oanaka, H and Okumura, T (1979)
"Engineering properties of lime treated marine soil and DM method"
Proc 6th Asian Reg Conf on Soil Mech and Found Engg, Singapore, Vol 1, pp 191–194

8.3   MIX-IN-PLACE BY SINGLE AUGER

Columns of soil-cement or lime stabilised soil can be created in the ground by the introduction of cement or lime paste through the hollow stem of an auger, the mixing taking place as the auger blades rotate. The auger is first drilled to the required depth of the treatment. It is then rotated and counter-rotated, raised and lowered a prescribed amount as the cement or lime is injected, to achieve consistent mixing.

Depending on the available torque of the rig and the required diameter, various soils can be strengthened by this technique from very weak clays, sludges, to sands and even firm or stiff clays.

The technique is available in the UK. Greenwood (1988) and Elliott (1989) explain the method used to stabilise alluvium (clayey near the surface becoming sand/gravel with cobbles) so that a 10 m-deep 10 m-diameter shaft could be constructed very close to existing buildings. Blackwell (1992) describes its use to form 14 m-deep stabilised walls of interlocking 0.75 m-diameter columns in sands for the construction of manholes in Rochdale. The rig was one used for cfa piling. The water-cement ratio of the injected material was 0.4, with typical injected volumes of 60 l/m of depth above the water table and 90 l/m below it.

Balossi Reselli *et al* (1991) describe the strengthening of a soft clay (s_u < 30 kN/m^2) in Italy by the formation of 1 m-diameter columns to 20 m.

Figure 8.7 shows a large-diameter auger (5 m in diameter) which, in sufficiently soft material, can be used to depths of 12 m.

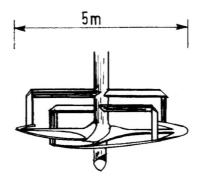

Figure 8.7   *Large-diameter mix-in-place auger*

The trade names used for this technique can be confusing. A US company describes this basic technique as "shallow mixing" in contrast to a "deep" multiple auger system capable of mixing to depths of 40–50 m. In the UK, the technique is termed "deep stabilisation" to contrast with the cement stabilisation of pavements.

8.3.1   References

Balossi Reselli, A, Bertero, M and Lodigiani, E (1991)
"Deep mixing technology to improve the bearing capacity of a very soft clayey soil under an earth embankment"
Proc 4th Int Conf on Deep Foundations, Stresa, Italy, April 1992

Blackwell, J (1992)
"A case history of soil stabilisation using the mix-in-place technique for the construction of deep manhole shafts at Rochdale"
Proc Int Conf on Grouting in the ground, Instn Civ Engrs, November 1992
Thomas Telford, London

Elliott, J A (1989)
"Deep mix-in-place soil/cement stabilisation using hollow stem augers for excavation at Elland, West Yorkshire"
In: *Piling and Deep Foundations* (J B Burland and J M Mitchell, eds)
AA Balkema, Rotterdam

Greenwood, D A (1988)
"Substructure techniques for excavation support"
Proc conf economic construction techniques – temporary works and their interaction with permanent works
Instn Civ Engrs, London

8.4   LIME STABILISATION OF SLOPES

Handy and Williams (1967) describe perhaps one of the first uses of lime to stabilise a landslide. Quicklime was introduced into 225 mm-diameter borings at 1.5 m centres, drilled to the base of the slide over an area of 75 m × 25 m. As part of the experiment two houses on the previously unstable slope were monitored. They virtually stopped moving after treatment, while adjacent houses continued to fail.

Another example of lime stabilisation was where sidelong embankments were built on unstable natural slopes of clay shales, weathered mudstones or phyllites in Thailand, reported by Reunkrairergsa and Pimsarn (1982) and Younger and Rananand (1985). One situation is shown in Figure 8.8 where holes were hand augered at about 3 m centres along the slope after it had been benched in steps. Borings were taken to the potential failure surface. Water and lime were added simultaneously and the filling operation was continued for one to two months as the lime migrated. The gain of *in-situ* strength of the soil with time was determined using a Borehole Direct Shear Device (Handy and Fox, 1967). It was noted that it was natural seepage beneath the bank that produced the migration of the lime.

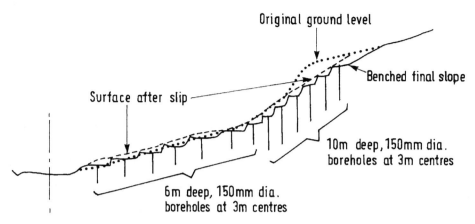

Figure 8.8   *Slope stabilisation by lime boreholes (after Younger and Ramanand, 1985)*

8.4.1   References

Handy, R L and Williams, W W (1967)
"Chemical stabilization of an active landslide"
Civil Engineering, Vol 37, No 8, pp 62–65

Handy, R L and Fox, N S (1967)
"A soil borehole direct shear test device"
Highway Research News, No 27, Spring, pp 42–51

Reunkrairergsa, T and Pimsarn, T (1982)
"Deep hole lime stabilisation for unstable clay shale embankment"
Proc 7th South East Asian Geotech Conf, Hong Kong, pp 631–645

Younger, J S and Rananand, N (1985)
"Ground improvement works, use of geotextiles and modern piling methods in Thailand"
Proc Symp on Recent developments in ground improvement techniques, Bangkok, 1982
AA Balkema, Rotterdam, pp 255–276

8.5   LIME STABILISATION OF PAVEMENTS

8.5.1   Definition

The engineering properties of clay subgrades, in particular plasticity, trafficability and strength, are improved by intermixing with lime and subsequent compaction. The surfaces of clay fills can similarly be stabilised with lime using mix-in-place methods to form capping layers prior to sub-base construction.

8.5.2   Principle

For pavement construction, the addition of lime to clay soils changes their character in the following ways (Bell and Tyrer, 1987; Bell 1988).

1. Lime and water accelerate the disintegration of clay clods during pulverisation. Soils therefore become friable and easy to work. However, the addition of the lime affects soils in different ways, depending on the type and length of curing and method of construction.

2. The lime absorbs moisture from the soil, in effect reducing its free moisture content. Maximum dry density decreases allowing an increase in optimum moisture content. After wet mixing, because of these changes, compaction should follow as soon as possible.

3. The plasticity of clay soils is changed by the addition of lime; plastic limits generally increase and liquid limits usually decrease. The plasticity index may fall by up to four times.

4. Linear shrinkage and swell properties decrease markedly. Changes in volume caused by variations in seasonal moisture contents in expansive or shrinkable clays can be minimised.

5. Unconfined compressive strengths increase greatly, perhaps by up to 40 times that of the original soil. This effect depends on temperature and will cease below 4°C.

6. California Bearing Ratio increases substantially.

7. Resistance to erosion is increased because of the binding effect of the lime. Resistance to frost action of siltier soils may be improved, but other soils may become more frost susceptible.

8. A lime-stabilised sub-base, capping layer or subgrade reduces penetration of water: when compacted it sheds water readily.

The reaction between clay and lime only proceeds when water is present, so it is essential to keep the mixed layer wet. Additional water may be required either from a bowser or spray and from a mixer head, the preferred method. Further guidance is given by Sherwood (1993) and the National Lime Association (1987) in their *Lime Stabilisation Construction Manual* describing practice in the United States.

8.5.3 Applications

Stabilisation of pavements on highly plastic clays using lime is a widespread practice in the United States, France and West Germany. In the UK, it was accepted rather later as a permitted material to form a pavement capping in the *Specification for Highway Works* (DOT, 1986). More powerful mix-in-place plant is available; for example, the Bomag MPH 100 Pulvimixer is reported to mix efficiently to a depth of about 350 mm. It is suggested that quicklime is more economical than hydrated lime because weight for weight there is about 25 per cent more available lime. Quicklime is a granular product that can be crushed and screened into different gradings. Hydrated lime is a much finer, lighter powder and dust can be a problem. Quicklime also has a faster drying action in wet soils, but has greater health and safety risks in use.

One unusual use of hydrated lime mixed with very soft clay is reported by Jones and Kim (1984). They describe how about 3 m of "Bay Mud" was treated to give an allowable bearing pressure of about 150 kN/m². In-service settlements of about 5 mm were measured.

8.5.4 Limitations

Adding lime in highly plastic clay soils can be highly beneficial. But Mitchell (1982) suggests that some organic compounds can retard or inhibit reaction. Free sulphates in soils can also cause the failure of a road as Mitchell (1986) describes. The sulphates deplete the available lime and together with expansive clay minerals, such as ettringite, can cause heaves of several centimetres. It is suggested by Mitchell (1986) that lime should not be used if the soluble sulphate content exceeds 50 000 parts/million.

Brandl (1982) also points out that soil/lime mixtures are often more frost-susceptible than untreated soil. Permeability can also increase with curing time (Wild *et al*, 1989).

8.5.5 Design

The DOT *Specification for Highway Works* defines the use of lime stabilisation for capping layers. Guidance on the design and use of lime stabilisation for pavements is given by Sherwood (1993) and Kennedy (1997) as well as in the *Lime Stabilisation Manual* (BACMI, 1988) and in the *Lime Stabilisation Construction Manual* (National Lime Association, 1987), which deal with UK and USA practices respectively. Useful papers are in the conference proceedings *Lime stabilization* (Rogers, Glendinning and Dixon, 1996).

8.5.6 References

Bell, F G (1988)
"Stabilisation and treatment of clay soils with lime"
"Part 1 – basic principles"
"Part 2 – some applications"
Ground Engineering, Vol 21, No 1, January, pp 10–15 (Pt 1) and
Vol 21, No 2, March, pp 22–30 (Pt 2)

Bell, F G and Tyrer, M J (1987)
"Lime stabilisation and clay mineralogy"
Proc Conf Foundations and Tunnels – '87
Engineering Technics Press, Edinburgh, Vol 2, pp 1–7

Brandl, H (1982)
"Alteration of soil parameters by stabilisation with lime"
Proc 10th Int Conf Soil Mech and Found Engg, Stockholm, 1981
AA Balkema, Rotterdam, Vol 3, pp 587–594

BACMI (British Aggregate Construction Materials Industry) (1988)
Lime stabilisation manual
BACMI, London

DOT (Department of Transport) (various dates)
Specification for highway works
HMSO, London

Jones, W F and Kim, J (1984)
"Treated San Francisco Bay Mud for building foundations"
In: *Piling and Ground Treatment*
Thomas Telford, pp 119–124

Kennedy, J (1997)
"European standards and research – developments and applications"
Recent application and benefits of lime treatment for earthworks and pavements
British Lime Association seminar, London, 15 January

Mitchell, J K (1982)
"Soil improvement: state-of-the-art report"
Proc 10th Int Conf Soil Mech and Found Engg, Stockholm, 1981
AA Balkema, Rotterdam, Vol 4, pp 509–565

Mitchell, J K (1986)
"Practical problems from supervising soil behaviour"
J Geotech Engg Div, Am Soc Civ Engrs, Vol 112, No 3, March

National Lime Association (USA) (1987)
Lime Stabilisation Construction Manual
National Lime Association, Washington, DC, 7th edn

Rogers, C D F, Glendinning, S and Dixon, N (1996)
Lime Stabilization
Thomas Telford, London

Sherwood, P T (1993)
Soil stabilization with cement and lime
TRL State of the Art Review, HMSO, London, 162 pp

Wild, S, Arabi, M, Leng-Ward, G (1989)
"Fabric development in lime treated clay soils"
Ground Engineering, Vol 22, No 3, April, pp 35–37

8.6 CEMENT STABILISATION OF PAVEMENTS

8.6.1 Definition

The engineering properties of suitable soil subgrades, particularly strength, stiffness and bearing capacity to traffic loading, are improved by intermixing with cement and subsequent compaction. The surfaces of earth fills of suitable characteristics can also be stabilised with cement using mix-in-place methods to form either capping layers prior to sub-base construction or, for lightly trafficked roads, the sub-base or base.

8.6.2 Principle

The US Highway Research Board defined soil-cement as "a hardened material formed by curing a mechanically compacted, intimate mixture of pulverised soil, cement and water" (HRB, 1961); with the additional explanation that the cement hydrates with the moisture in (or added to) the soil to achieve the hardening. This succinct definition explains the principle of cement stabilisation. The key steps are:

- pulverisation of the soil so that it can receive and be mixed with cement
- thorough intermixing of cement throughout the soil
- addition of water, primarily for compaction but which is also available to the cement
- compaction.

The suitability of the soil for cement stabilisation depends largely on its plasticity, particle size distribution and organic content. Increased strength and stiffness are achieved by the creation of a more rigid soil structure as in Figure 8.1.

8.6.3 Description

Plant-mix methods are available, in which the soil is brought to a central plant for drying, pulverisation, mixing and water content control. As with lime stabilisation, the development of more powerful mix-in-place equipment, which can either be multi-pass or single pass, is extending the use of this technique. The mix-in-place operations are of rotovation to pulverise the soil, mixing the cement into the soil, mixing additional water, and spreading the material uniformly prior to compaction. Kennedy (1983) describes the methods and usages of cement stabilisation.

Comprehensive reviews of cement-stabilised bases and sub-bases are given in a 1972 symposium (Institution of Highway Engineers, 1972). The publication with particular authority is that by Sherwood (1993) in his TRL state-of-the-art review.

Depending on soil type, adequate stabilisation for lightly trafficked road bases can be achieved with cement contents of 3–6 per cent (Charman, 1988). Higher proportions tend to be uneconomic; lower contents are unlikely to be mixed well enough. Higher proportions of cement are needed for poorly graded and finer grained soils (see Table 8.2) and when the purpose is to produce a cement-bound material for a major road (rather than stabilising an *in-situ* soil).

Typical depths of treatment are 150–200 mm (ie the finished, compacted layer). Although more powerful, heavier plant can now cope with thicker layers, the critical operation is then ensuring adequate compaction.

Where the soil is a plastic (heavy) clay, it can be rendered more workable (ie friable) by mixing it with 2–3 per cent of lime before cement stabilisation.

Table 8.2 *Approximate cement contents for materials capable of being stabilised to produce cement-bound roadbases and sub-bases (from Kennedy, 1983)*

Material type	Approximate cement requirement
Well-graded hard granular materials	80–120 kg/m^3
Pulverised fuel ash	80–240 kg/m^3
Poorly graded/uniform, hard granular materials and well-graded, weak rocks such as shale	130–190 kg/m^3
Brickearths	170–200 kg/m^3
Soft chalk	180–225 kg/m^3

8.6.4 Applications

Soil-cement is widely used abroad, particularly for low-cost rural roads in Africa, South America, USA and Australia. It is still used for landing strips, continuing the use made of this technique by military engineers in the Second World War. As well as speed of construction, the main advantage is the use of locally available material (eg see Charman, 1988). In UK, relatively few soil-cement roads have been built, although cement-bound sub-bases and cappings are permitted (DOT, 1986).

Soils stabilised with cement have been used in various other ways, particularly as linings for canals and irrigation ditches and even as a rip-rap substitute, but these usages are as a low-cost construction material rather than for ground improvement.

8.6.5	**Limitations**

The main limitations are: the shallow depth of treatment; frost susceptibility; and the development of shrinkage cracks. Highly organic soils cannot be treated successfully with cement.

8.6.6	**Design**

Guidance on material selection, mix design, construction and control testing is given by Sherwood (1993), Kennedy (1983), Charman (1988) and Ingles and Metcalf (1972). The use of cement stabilisation for capping layers is covered in the DOT *Specification for Highway Works*.

8.6.7	**References**

Charman, J H (1988)
Laterites in road pavements
Special Publication 47, CIRIA, London

DOT (Department of Transport) (various dates)
Specification for highway works
HMSO, London

Highway Research Board (HRB) (1961)
Cement-stabilised soil
HRB Bulletin 292, Washington, DC

Ingles, O G and Metcalf, J B (1972)
Soil stabilisation
Butterworth, Sydney, Australia

Institution of Highway Engineers (1972)
"Cement stabilized bases and sub-bases, Parts 1 and 2" (seven papers)
J Instn High Engrs, February and March

Kennedy, J (1983)
Cement-based materials for subbases and road bases
Cement and Concrete AsSoc, Wexham Springs

Sherwood, P T (1993)
Soil stabilization with cement and lime
TRL State of the Art Review, HMSO, London, 162 pp

9 Improvement by grouting

A general definition of grouting for ground improvement is: "the controlled injection of material, usually in a fluid phase, into soil or rock in order to improve the physical characteristics of the ground". Such a definition does not cover all types or purposes of grouting in the ground, eg grouting to raise ground slabs or road pavements, but it does cover grouting to fill voids in the ground, whether natural (such as in karstic limestone) or resulting from human activity.

CIRIA has recently completed a project to prepare guidance documents on the subject of grouting, so the following sections are introductory in nature. One of the first stages of the project was the issue of a *Glossary of terms and definitions used in grouting: proposed definitions and preferred usage* (Project Report 61, WS Atkins, 1997), following issue of a draft by Sherwood (1991) for industry comment. A grouting bibliography has also been prepared (Project Report 60, Hellawell *et al*, 1997). Another of the base documents for this project was the explanation of the underlying engineering science in grouting by D A Greenwood (1997, although first issued in 1992). Detailed guidance on techniques, materials, plant and the contractual frameworks of grouting is given in CIRIA Publication C514 (Rawlings *et al*, 1999).

9.1 GROUTING PROCESSES

The wide scope of grouting works, their historical development and variations in terminology, which is often associated with proprietary processes, have led to alternative classifications of types of grouting. Indeed, as techniques and their usages have changed over the years, so have the terms and the proposed classifications.

Greenwood (1997) has proposed that grouting methods be classified by process rather than by grout material or end result. He considers four main processes involving grout injection:

- permeation
- hydrofracture
- mechanical mixing
- ground compaction.

To these may be added void filling (see Figure 9.1).

These can be further subdivided by process variations. Thus mechanical mixing can be divided into (1) mix-in-place and (2) some forms of jet grouting. The first of these is covered in Chapter 8 (Improvement by admixtures) and specifically in Sections 8.1–8.3. It is arguable whether jet grouting should properly be classified with mix-in-place treatments, as it totally destroys the original soil fabric, remixing or replacing it. Some systems of jet grouting result in a stabilised column consisting of a mixture of soil and grout. Other systems produce columns in the ground by the total replacement of the soil with grout, ie it is not a mixing process. These replacement processes could be thought of as improvement by structural reinforcement (as could the stone columns of vibro-replacement). However, both the replacement processes and mixing systems of jet grouting are included in this section of the report because the term "grouting" in the name is widely accepted and the injected materials can be considered to be grouts.

The definitions and principles of improving the ground by grouting processes (illustrated in Figure 9.1) are presented in the following section:

- permeation grouting (9.2)
- hydrofracture grouting (9.3)
- jet grouting (9.4)
- compaction, squeeze and compensation grouting (9.5)
- cavity infilling (9.6).

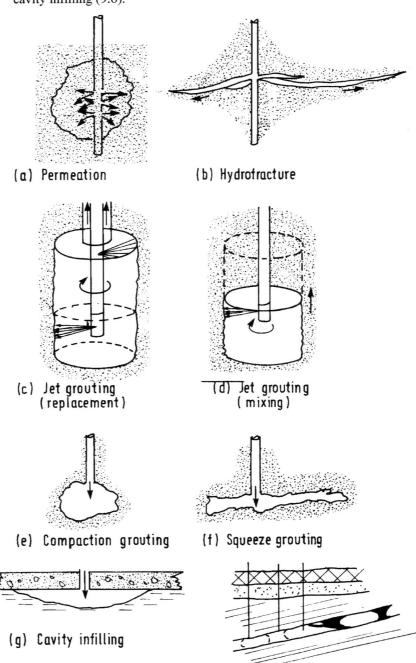

(a) Permeation

(b) Hydrofracture

(c) Jet grouting (replacement)

(d) Jet grouting (mixing)

(e) Compaction grouting

(f) Squeeze grouting

(g) Cavity infilling

Figure 9.1   *Main processes of grouting*

As the descriptions of the methods and their usages have many common aspects. To set them in context, the following sections deal with these general matters:

- the reasons for grouting (9.1.1)
- assessment of groutability (9.1.2)
- types of grout (9.1.3)
- plant and equipment (9.1.4)
- controls (9.1.5).

As may be seen from the wide range of techniques classed as grouting, the selection of an appropriate one to a particular set of ground conditions and purposes requires experience and judgement. Equally important is that to be successful each grouting project should be set up with an observational approach so that treatments can be modified or adjusted as the work proceeds

9.1.1 The reasons for grouting

There are nine main reasons for using grouting as the means of ground improvement.

1. Reducing the permeability of the ground.

2. Providing a barrier or cut-off to water flow in the ground.

3. Increasing the strength of the ground to permit its safe excavation.

4. Increasing the strength, stiffness and bearing capacity of the ground.

5. Increasing the overall density of the ground.

6. Infilling natural cavities.

7. Infilling mineral workings.

8. Infilling voids adjacent to structures.

9. Compensating for ground loss caused by adjacent works.

The above list shows the versatility of grouting techniques, made possible by the wide ranges of methods, grouting materials and the scope for using them in combination.

9.1.2 The assessment of groutability

Grouting, for whatever purpose, affects the void spaces of the ground either to fill them or to reduce their volume. Even in the mixing and replacement processes of jet grouting the proportion of voids in the *in-situ* soil influences the ease with which the jetting erodes the soil and the quality of mixing.

The groutability of the ground is not a single or fixed characteristic, but depends on the purpose of the grouting, the technique and the grout as well as the nature of the ground and its voids. Different techniques or combinations of techniques with different grouts can achieve the same purpose. Very broadly, however, the grouting processes associated with different types of ground are shown in Table 9.1.

Table 9.1   *Grouting processes associated with different types of ground*

Type of ground	Grouting process				
	Permeation	**Hydrofracture**	**Jet grouting**	**Compaction and squeeze grouting**	**Infilling**
Gravel	✓	—	—	—	—
Sand	✓	✓	✓	✓	—
Silt	—	✓	✓	✓	—
Clay	—	✓	✓	✓	—
Decomposed rock	✓	✓	✓	✓	—
Fissured rock	✓	✓	—	✓	—
Voided ground	—	—	—	✓	✓
Fills	—	✓	✓	✓	—

From this it can be seen that the processes that use a thick paste or mortar type of grout have wide application, ie void infilling and the compaction/squeeze grouting methods. Jet grouting also can be carried out in most soils – it would not usually be appropriate in gravels or in bouldery soils. Hydrofracture grouting is used for both soils and rocks. The assessment of groutability therefore refers largely to permeation grouting, ie the ability of the pores or fissures to be permeated by different types of grout.

The pore/fissure structure of ground is usually characterised by soil particle size distribution, typical fissure sizes or by equivalent coefficient of permeability.

Various authors present this type of information in different ways and not always consistently. Figure 9.2 presents, for different types of grout, the limits of permeation suggested by various authors in terms of soil grading and hydraulic conductivity (coefficient of permeability to water). While there is broad agreement on gradings, there are differences when comparing in terms of permeability. The closer the soil is to the finer limit of size the greater is the chance of the onset of hydrofracture.

One of the critical factors, not made sufficiently clear by the grading curves of Figure 9.2, is the control that a small difference in fines content can make – because of the way small particles fill and occlude the larger pores between coarse particles. Baker (1982) illustrates this in relation to permeation by chemical grouts (see Figure 9.3).

The need for rock to be grouted is normally assessed by packer permeability tests carried out in a series of tests at increasing and then decreasing pressure heads of water. On their own, these tests are not sufficient to indicate the groutability of the rock, but their results have to be assessed in relation to the discontinuity and rock fabric characteristics (see Ewert, 1994, for a discussion of this).

Mitchell and Katti (1981) proposed the empirical "groutability" ratios for particulate grouts in soils and fissured rocks shown in Table 9.2.

9.1.3   Types of grout

Numerous materials have been developed for use as grouts. All have the property of fluidity, but this can be of materials as diverse as thick pastes and mortars, slurries, colloidal suspensions in water, emulsions of one liquid in another, solutions, liquid resins and, even, suspensions of very fine solids in gases. As with the long-established Joosten process, which involves two phases of injection (sodium silicate first and a reagent, calcium chloride, subsequently), a grout is not necessarily a single material.

Table 9.2   *Groutability of particulate grouts in soils and fissured rocks (after Mitchell and Katti, 1981)*

For soils: $N = \dfrac{(D_{15})_{soil}}{(D_{85})_{grout}}$

where $N > 24$ grouting consistently possible, but where $N < 11$ grouting is not possible

or $N_c = \dfrac{(D_{10})_{soil}}{(D_{95})_{grout}}$

where $N_c > 11$ grouting consistently possible, but where $N_c < 6$ grouting is not possible

For rock: $N_R = \dfrac{(\text{width of fissure})}{(D_{95})_{grout}}$

If $N_R > 5$ grouting is consistently possible, but if $N_R < 2$ grouting is not possible

With such a range of materials, various classification systems have been proposed on the basis of, for example, their constituent materials, the nature of the grout fluid or the setting process, and the rheological characteristics of the grout. It is convenient, however, to consider the main types of grout as follows:

- mortars and pastes (usually a combination of cementitious material and filler)
- cement slurries and suspensions in water
- cement-bentonite suspensions in water
- bentonite (or other clay) colloidal suspension in water
- microfine cement colloidal suspensions in water
- chemical solutions usually in water
- chemicals in a colloidal system that form gels
- resins, ie organic chemicals.

As can be seen from Figure 9.2, these grout types, in descending order of relevance to permeation, are associated with increasingly finer pore spaces. The thicker grouts are used for void filling, jet grouting, compaction and squeeze grouting. The intermediate and colloidal suspensions are often used for hydrofracture.

Cement-based grouts

These probably remain the most widely used grouts. Rock fissure grouting is often entirely based on using slurries of cement in water, at proportions of, say, 1:30 or 1:15 initially, with subsequent injections at higher proportions of cement until the "take" is acceptably low. It is normal to use a bentonite suspension (of 2–5 per cent dry weight of bentonite) to carry the cement particles in suspension, and to improve the mobility of the grout and its impermeability when set.

An important development is the manufacturing of very fine-grained cements, referred to by names such as microfine, superfine and ultrafine cement. Ordinary Portland (OP) cements have gradings typically with 10 per cent of their particles coarser than 50 μm and only 10 per cent finer than 5 μm in size. It is because of the coarser particles that OP cements can only be used to penetrate coarse-grained soils and open fissures. The microfine cement gradings, by contrast, have a mean particle size of about 5 μm and a maximum of 10–15 μm.

The mixes used for jet grouting are relatively thick slurries using proportions of water to cement of 0.5 or 0.6 to 1. Other mixtures, eg with 2–5 per cent of bentonite and/or the addition of pfa, are also used (Bell and Burke, 1994). The mixes are designed to achieve the target strength needed for the column formed in the ground, but in relation to the requirements of the jetting process.

For compaction grouting, mortar pastes have typical slumps of 50–75 mm, and consist of mixtures of cement, sand and water, often with pfa or other cementitious material and a small proportion of bentonite.

Cement-bentonite mixtures of various proportions and stiffnesses are often used for hydrofracture grouting when the purpose is to create a tracery of impermeable sheets or lenses in the ground.

Chemical grouts

This term is often used to mean grouts that are solutions, or where the chemical is a liquid that contains no particles in suspension. Thus resins are sometimes considered as chemical grouts and sometimes as a separate class.

Many of the chemical grouts are based on the combination of sodium silicate and a reagent to form a gel. The Joosten process used in coarse granular soils uses calcium chloride as the reagent. Other reagents are organic esters, sodium aluminate, and bicarbonates. The reagent and the proportions can be chosen to control the gel time, the initial viscosity and the order of strength of the grouted soil.

Phenolic, acrylic and amino resins are used for permeating fine-grained soils, the choice depending upon the desired viscosity, gel time and strength.

Table 9.3, from Greenwood and Thomson (1984) and Karol (1990) summarises typical characteristics of chemical grouts.

Table 9.3   *Typical characteristics of chemical grouts*

Grout	Initial viscosity (range, centipoise)	Strength	Gel times (min)	Corrosivity or toxicity
Silicates	Low to high (1.5–40)	Low to high	1–200	Low to medium
Lignin compounds	Low to medium (2.5–20)	Low	5–120	High
Phenolic resins	Low to medium (1.5–10)	Low	5–60	Medium
Acrylic resins	Low to medium (1.3–10)	Low	1–200	High
Amino resins	Medium to high (6–30)	High	40–300	Medium
Acrylates	High	Low	—	Low
Polyurethanes	High (19–150)	High	React instantly with water	High

All gelling chemicals are toxic if mishandled; some grouts are no longer permitted or not in certain situations. Their use is subject to the COSHH Regulations (Control of substances hazardous to health). Environmental regulations to prevent pollution may also limit the use of chemical grouts.

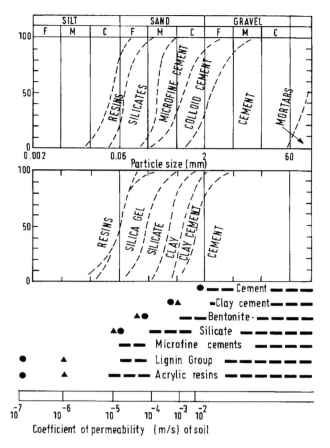

Note: circle and triangle symbols represent additional lower bounds suggested by the authors referenced below

Figure 9.2   *Groutability for permeation in terms of grain size and coefficient of permeability of soil (after Tausch, 1985; Schlosser and Juran, 1981; Karol, 1990; Tornaghi, 1980)*

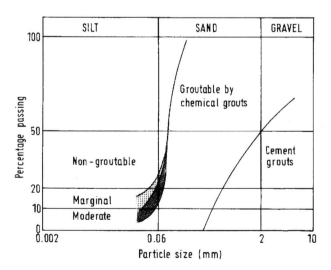

Figure 9.3   *Groutable range for chemical grouts (after Baker, 1982)*

Sherwood and Gandais (1994) describe a grout designed to be compatible with stringent environmental conditions that preclude the use of organic chemical grouts. It combines extremely fine, pure silica and calcium hydroxide, which promotes a pozzolanic cementation. The particles in the grout mixture have sizes largely finer than 2 μm (mean size about 0.7 μm).

Other types of grout

For filling large natural cavities or old mine workings, the infill grouts are often based on cheap, locally available materials, eg minestone or similar wastes. They may or may not include a cementitious material, whether cement, pfa, slag or pozzalans, but because of the large volumes involved these are kept to a minimum to reduce cost. The requirements for quality control, while not as onerous as for other forms of grouting, are not dissimilar in that the grout mix has to be readily pumpable to the target locations, but then become sufficiently rigid while so far as possible maintaining its volume fully in the void space.

9.1.4 Plant and equipment

All grouting operations involve two essential items of plant, an adequate mixer and an appropriate pump and three essential sets of equipment: the distribution system, the control/monitoring apparatus and the injection points. Usually there will be additional plant or facilities for the handling and storage of the grout materials, for the agitation of mixed grout before pumping, and for the removal and disposal of spoil and waste grout.

With the advent of electronic transducers and micro-processors, integrated systems are now used for batch control (mix proportioning, volume for injection) and pumping (pressure limit definition) and for recording these parameters for each injection point. Such systems not only improve the control and, in effect, workmanship of individual injections, but they also provide the grouting engineer with the comprehensive, reliable information needed to adjust the grouting to best effect (see Greenwood *et al*, 1987; Guillaud and Hamelin, 1994; Harris and Cottet, 1994).

Mixers

Grouts that are slurries or colloidal suspensions must be mixed with high-shear mixers, ie the fluid is repeatedly subjected to high shear forces to achieve the mixing. Although these involve high-speed rotors or rollers, the process depends on the shearing rather than the rapid movement of the fluid. Three types of high-shear mixers are shown in Figure 9.4. These contrast with pan-and-paddle mixers (Figure 9.5), which are used for agitation of mixed colloidal grouts or for simple cement-water slurries.

Pumps

Greenwood (1997) notes that the choice of grout pump must be accord with the intended technique. "All too frequently the technique is dictated by the pump characteristics rather than the desirable grout properties." Many ram or piston-type pumps develop peaky pressures, which can lead to unintended hydrofracture. The choice of pump, therefore, has to suit the following aspects, depending on the purpose of the grouting:

- nature of the grout (eg viscosity, abrasiveness, particle sizes etc)
- required volume flow rate, which may be very low for resin injection to relatively high for void filling
- required control at a single injection point, which might be as little as a few litres
- required pressure and pressure limits.

A wide range of pumps is available of both the progressive cavity and positive displacement (piston) types. The ability to monitor and control pressures and volume flow rates electronically is a major advance over manual control and pressure gauge observation.

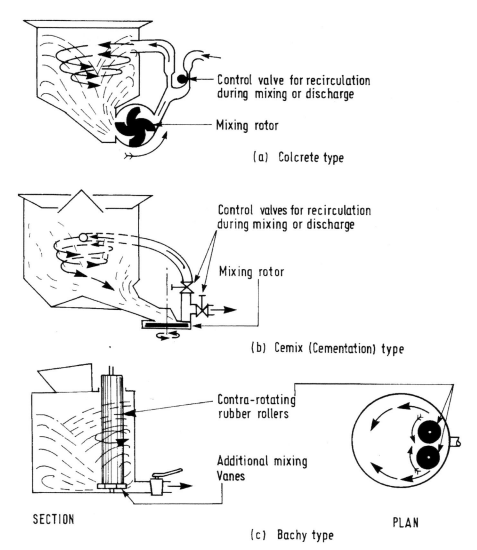

Control valve for recirculation during mixing or discharge

Mixing rotor

(a) Colcrete type

Control valves for recirculation during mixing or discharge

Mixing rotor

(b) Cemix (Cementation) type

Contra-rotating rubber rollers

Additional mixing Vanes

SECTION

PLAN

(c) Bachy type

Figure 9.4 *Three types of high-shear grout mixers*

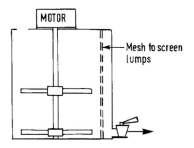

MOTOR

Mesh to screen lumps

Figure 9.5 *Pan-and-paddle mixer for agitation or mixing simple slurries*

Distribution

The distribution system has to be integrated with the mixing and pumping plant and with the control and monitoring methods. Among other matters necessary to achieve the required quality, the distribution system has to minimise:

- pumping lengths

- pressure losses in transmission

- segregation of the grout mixture or other loss of quality

- wastage on transference from one injection point to another and at cleaning out.

Injection points

Various options are available depending on the type and purpose of the grouting. The three basic methods are as follows.

1. Drillholes using packers (see Figure 9.6).

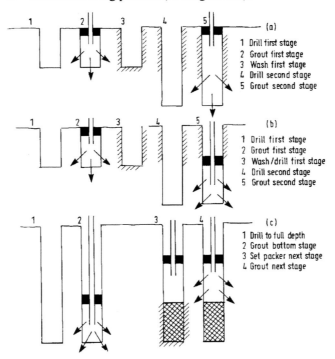

1 Drill first stage
2 Grout first stage
3 Wash first stage
4 Drill second stage
5 Grout second stage

1 Drill first stage
2 Grout first stage
3 Wash/drill first stage
4 Drill second stage
5 Grout second stage

1 Drill to full depth
2 Grout bottom stage
3 Set packer next stage
4 Grout next stage

Figure 9.6   *Drilling and grouting stages in rock fissure grouting:*
(a) downstage with top packer, (b) descending stages,
(c) ascending stages

2. Lances, ie driven tubes, perforated over part of their length, which may be used for injection as they are driven in or withdrawn (or in both stages, as in the Joosten process) or with an internal tube and packer system. These are only suitable for shallow depths of treatment.

3. *Tubes à manchette* (TAM) (see Figure 9.7) ie the installation in a drillhole, of an access tube with rubber-sleeved (covered) perforations at known depths. A grout pipe with a perforated section between packers is installed in the access tube and positioned at the required injection depth. In the basic system the access tube is set into the drillhole in a weak, usually cement-bentonite, mixture called sleeve grout. The first step of injecting at any point therefore involves displacement or even rupture of the rubber sleeve and hydrofracture of the sleeve grout in order for the injection grout to reach the ground to be grouted.

    The advantages of the TAM method are of injection at specific depths over the zone to be treated and the capability of both multiple injections at a point and of staged injections, ie primary, secondary depth intervals. The packer shown in Figure 9.7 is rather dated, but shows the principle. Modern packers are inflatable with flexible grout pipes. A development of the TAM approach is the multiple-sleeved pipe system (Bruce, 1991) in which the access tube is stabilised in the drillhole by fabric collars inflated in place with cement grout, ie there is no sleeve grout. The access tube is perforated and sleeved as with the basic TAM method. After the initial injection between two of the inflated collars, that part of annulus between the access tube and the surrounding ground is filled with grout. This can make for differences in the sequencing of injections compared with the basic TAM method.

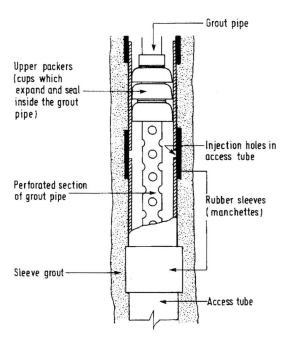

Figure 9.7 *The* tube à manchette *system (after Bruce, 1982)*

A critical part of the design of grouting works is the efficient arrangement of the injection points, ie their installation in the ground to permit effective treatment of all the ground necessary to create the three-dimensional grouted shape. This is often complicated by limited access or because of the need to avoid underground services, obstructions or existing structures.

9.1.5 Controls

The whole point of grouting is to place the correct grout in the correct position in the ground. Knowing where the grout actually goes in the ground is often difficult, however. Hence control system development concentrates on getting the right grout placed in the correct position at the injection point, in a system that combines observation and further treatment. There are too many variations of technique and materials to be specific in this report, but in general terms the control of grouting includes:

- quality of grout materials and their storage and handling
- quality of the mixed grouts and rheological properties in respect of storage, pumpability and setting
- quality and durability of the set grout
- cleanliness of pumping lines and other equipment
- accuracy of batching systems for mixing and for injection, including where two liquids are mixed in the line
- accuracy of pressure control systems for pumping and injection and compatibility between the pump type and the required pressure and flow rate for the type of grouting
- drilling and accurate positioning of injection points
- observation of developing "tightness" of the ground to new injections and of ground movements
- data-logging of records and on-site control by visualisation or plotting of monitoring results
- use of a system that allows repeat or additional injections.

Testing the effectiveness of the treatment depends on its purpose. *In-situ* permeability tests may be adequate to check a cut-off, but in some cases a pumping test might be desirable. Observations of ground heave or building movements by precise levelling might be appropriate in other situations, such as with compensation grouting. Load tests on treated ground would be used where jet grout columns form part of a foundation.

9.1.6 References

WS Atkins Consultants (1997)
Glossary of terms and definitions used in grouting: proposed definitions and preferred usage
Project Report 61, CIRIA, London

Baker, W H (1982)
"Engineering practice of chemical grouting"
In: *Improved design and control of chemical grouting*
Report FHWA/RD-82-038, US Dept of Transport, Washington, DC, Vol 3

Bell, A L and Burke, J (1994)
"The compressive strength of ground treated using triple system jet grouting"
In: *Grouting in the ground* (A L Bell, ed), Proc Int Conf at Instn Civ Engrs, Nov 1992
Thomas Telford, London, pp 525–538

Bruce, D A (1982)
"Aspects of rock grouting practice of British dams"
Proc ASCE Conf on Grouting in Geotechnical Engineering, New Orleans, pp 301–316

Bruce, D A (1991)
"Grouting with the MPSP method at Kidd Creek Mines, Ontario"
Ground Engineering, Vol 24, No 8, October, pp 26–41

Ewert, F-K (1994)
"Evaluation and interpretation of water pressure tests"
In: *Grouting in the ground* (A L Bell, ed), Proc Int Conf at Instn Civ Engrs, Nov 1992
Thomas Telford, London, pp 141–162

Greenwood, D A (1987a)
"Underpinning by grouting"
Ground Engineering, Vol 20, No 3, April, pp 21–32

Greenwood, D A (1997)
Fundamental basis of grout injection for ground treatment
Project Report 62, CIRIA, London. Also issued as draft document for CIRIA project, "The use of grouting techniques for ground improvement", January 1992

Greenwood, D A and Thomson, G H (1984)
Ground stabilisation
Thomas Telford, London

Greenwood, D A, Hutchinson, M T and Rooke, J (1987)
"Chemical injection to stabilise water-logged sand during tunnel construction for Cairo Waste Water Project"
Proc 9th Eur Conf Soil Mech and Found Engg, Dublin, 1987
AA Balkema, Rotterdam, Vol 1, pp 157–164

Guillaud, M and Hamelin, J-P (1994)
"Computerised grouting design and control"
In: *Grouting in the ground* (A L Bell, ed), Proc Int Conf at Instn Civ Engrs, Nov 1992
Thomas Telford, London, pp 215–226

Harris, R R W and Cottet, P (1994)
"Evaluation of computer control and analysis of grouting parameters during grouting works"
In: *Grouting in the ground* (A L Bell, ed), Proc Int Conf at Instn Civ Engrs, Nov 1992
Thomas Telford, London, pp 237–246

Hellawell, E E, Rawlings, C G and Kilkenny, W M (1997)
Geotechnical grouting – a bibliography
Project Report 60, CIRIA, London, 93 pp

Karol, R H (1990)
Chemical Grouting
Marcel Dekker Inc, 2nd edn, 465 pp

Mitchell, J K and Katti, R K (1981)
"Soil improvement: general report"
Proc 10th Int Conf Soil Mech and Found Engg, Stockholm, 1981
AA Balkema, Rotterdam, Vol 4, pp 567–578

Rawlings, G, Hellawell, E E and Kilkenny, W M (1999)
Grouting for ground engineering
Publication C514, CIRIA, London

Schlosser, F and Juran, I (1981)
"Design parameters for artificially improved soils"
In: *Design parameters in geotechnical engineering*
Proc 7th Eur Conf Soil Mech and Found Engg, Brighton, 1979
Brit Geotech Soc, London, Vol 5, pp 197–226

Sherwood, D E (1991)
Glossary of terms and definitions used in grouting: proposed definitions and preferred usage (draft)
Draft document for CIRIA project "The use of grouting techniques for ground improvement", October 1991, CIRIA, London, 24 pp

Sherwood, D E and Gandais, M (1994)
"Neutral mineral micro particle grout and its application in recovery of a collapsed tunnel in Switzerland"
In: *Grouting in the ground* (A L Bell, ed), Proc Int Conf at Instn Civ Engrs, Nov 1992
Thomas Telford, London, pp 631–648

Tausch, N (1985)
"A special grouting method to construct horizontal membranes"
In: *Recent developments in ground improvement techniques*
Proc Int Symp, Bangkok, 1982

Tornaghi, R (1980)
"Stabilization by injection – panellist report and discussion"
Design parameters in geotechnical engineering, Proc 7th Eur Conf Soil Mech and Found Engg, Brighton, 1979, British Geotech Soc, Vol 4, Sections 8.3 and 8.36

9.2 PERMEATION GROUTING

9.2.1 Definition

Permeation grouting is replacement of the water in voids between soil grains or in rock fissures with a grout fluid at an injection pressure low enough not to cause a fracture of the soil fabric or displacement of the rock blocks (Figure 9.1a).

9.2.2 Principle

The grout injection aims to fill fine pore spaces either to decrease the permeability or to increase the strength of a specific volume of the ground. Thus the grout has to be designed to achieve an optimum set of characteristics:

- minimal viscosity to flow the target distance from the injection point through tortuous, narrow passages

- sufficient fineness if it is a particulate grout so as not to create blockages too close to the injection point by the filtration of particles on pore openings

- a low enough early strength so as not to preclude a later grout stream during the injection from cutting through it

- an adequate strength and durability in place to provide the long-term improvement required.

For the degree of control implied by the above characteristics, the nature and structure of the ground has to be understood to inform choices about the location of injection points, the target volume of grout to be injected and the pressure that may be applied.

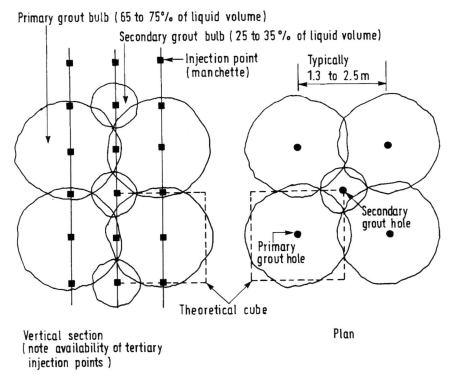

Figure 9.8 *Grout hole arrangement and injected volumes (after Baker, 1982)*

9.2.3 Permeation grouting: description and applications

Permeation grouting involves multiple injections, sometimes repeated at the same injection point, in order to create the required shape of treated ground. For this reason *tubes à manchette* and, in some circumstances, lances are used. The grouts are designed in terms of their flow and setting characteristics to match the pore sizes of the ground to be treated and the properties needed for the zone of treatment.

Permeation of a block of ground is an iterative process and the treatment is usually made by a sequence of injections first at an open spacing and subsequently at intermediate positions. In plan this may be from primary, secondary and tertiary groutholes; in elevation, from alternate or more widely spaced *manchettes*. Figure 9.8 shows a possible arrangement of grout holes and injection points.

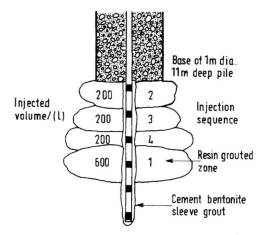

Figure 9.9   *Example of a treatment sequence to strengthen ground below a pile (after Littlejohn* et al, *1984)*

It is usual to grout the perimeter, base and top of the intended zone of treatment first, in effect to contain the subsequent grouting within the zone. An example showing the bottom-top-middle sequence is given in Figure 9.9 for grouting below the base of 11 m-long piles with resin grout to strengthen the coarse sands and interbedded silts.

As the purpose is filling the pore spaces an estimate is made of the pore volume of the ground in relation to the spacing of injection points and the rheological characteristics of the grout. Thus the permeation is designed and controlled in relation to these aspects, ie:

* knowledge of injection position
* injecting a predetermined volume of grout at that position
* controlling and recording the injection pressure
* controlling and recording the flow rate
* selecting subsequent injection positions and the control parameters in response to the treatment already carried out.

When the grouting scheme involves hundreds of separate injections the benefits of computer-based control and recording systems are obvious.

The two main purposes of permeation grouting are for the control of groundwater and for strengthening the ground, often together and to assist excavation. The grouting can be carried out either from the ground surface or, if this involves drilling too deeply, from within the excavation itself. Figure 9.10 shows a scheme of treatment to provide a stabilised and impermeable cover above twin tunnels where it was possible to work from the ground surface despite the presence of services. Figure 9.11, on the other hand, shows grouting at two levels in sinking a deep mine-shaft, where the grouting was needed for the construction of the shaft itself. Where a tunnel and its associated shafts are in water-bearing ground, grouting may not be needed specifically for constructing the shafts but for the break-out of the shaft side at the start or completion of the tunnel drive (see Coe *et al*, 1994 for a description of grouting for the Cairo Waste Water Project tunnels).

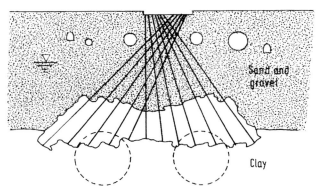

Figure 9.10 *Grouted zone prior to tunnel excavation: combined permeation of sand and gravel and hydrofracture of clay*

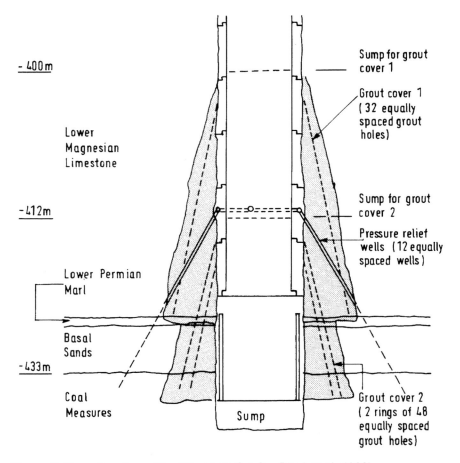

Figure 9.11 *Grouting of the Riccall shaft (after Black* et al, *1982)*

Permeation grouting is also used for underpinning. Figure 9.12 is an example of the patterns of injection that might be needed, depending on the position of the zone requiring treatment in relation to structures or obstructions.

The grouting of rock fissures is often associated with the construction of dams, eg to form a grout curtain below the dam. While fissure grouting can be achieved by permeation methods, ie the fissures being the targeted pore spaces rather than the pores within the intact rock, the techniques that have often been used are of cement injection and high pressures and are less clearly permeation. This can be more a matter of blocking the discontinuities with the cement particles, perhaps using high pressure as a means of widening the fissures. The finer cements and fine-grained silicas now available mean that this type of fissure grouting can be more effective.

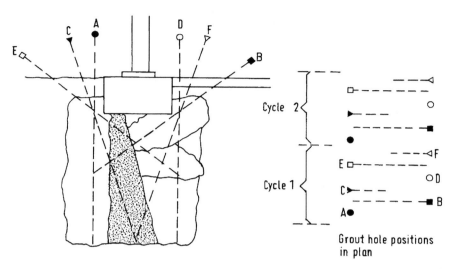

Figure 9.12 *A pattern of injection holes for underpinning by permeation (after Greenwood, 1987a)*

9.2.4   Limitations

As well as the questions of the groutability of the ground (see Section 9.1.2) and of positioning the grout injection locations, the performance of the grout while a fluid and its durability in place are also possible limitations. Karol (1982a) and Hewlett and Hutchinson (1983) point out that some chemical grouts can deteriorate in repeated freeze/thaw or wet/dry cycles and may even dissolve in prolonged contact with groundwater. Mongilardi and Tornaghi (1986) explain that syneresis, or shrinkage with water loss, can also occur. Groundwater with high pH conditions can accelerate the setting of sodium silicate grouts and prevent the setting of acrylamide or acrylate grouts (Baker, 1982). Organic materials in groundwater can also affect the gelling of chemical grouts.

Toxicity is an important aspect with chemical grouts, not only during site operations, but also because of the potential effects on any aquifer in contact with the grout treatment. Table 9.3 indicates rough levels of risk, of toxicity and corrosivity for principal groups of chemical grouts. Karol (1982) gives a detailed history of the development and then banning of the use of various grouts. He concluded that the ideal product was not yet available. Mongilardi and Tornaghi (1986) suggest the return to silica-based grouts, with inorganic reagents, that are as stable as cement grouts and are acceptable environmentally. Skipp and Hall (1982) give further guidance on the health hazards in relation to grouting operations. Cement grouts may be attacked by sulphates or highly acidic groundwater. Cement/bentonite grouts may collapse in contact with salt water (Mitchell, 1982).

9.2.5   Design

Many types of chemical grouts have been developed in recent years and they are described in detail in the book by Karol (1982a). Further guidance on grout character and behaviour, groutability, and grouting-programme design is given by Karol (1982b) and by Baker (1982). Injection concepts, grout zone geometry, estimation of grout volumes, grout pipe layouts/installation, injection staging/sequencing and monitoring are all discussed by Baker (1982).

The paper by Littlejohn (1982) discusses the design of cement-based grouts, and examines the importance of bleeding, fluidity, setting, shrinking, thermal properties, strength, and durability. How to minimise water content so that bleeding can be better controlled and durability improved is also discussed by Houlsby (1985).

9.2.6   References

Baker, W H (1982)
"Engineering practice in chemical grouting"
In: *Improved design and control of chemical grouting*
Report FHWA/RD-82-038, US Dept of Transport, Washington DC, Vol 3

Black, J C, Pollard, C A and Daw, G P (1982)
"Hydrogeological assessment and grouting at Selby"
Proc Conf Grouting in Geotechnical Engineering, New Orleans
Am Soc Civ Engrs, pp 665–679

Coe, R H, Greenwood, D A, Kleina, R and McAnally, C J (1994)
"A review of grouting on the Greater Cairo Wastewater Project"
In: Bell, A L (ed) *Grouting in the ground*, Proc Int Conf at Instn Civ Engrs, Nov 1992
Thomas Telford, London, pp 163–182

Greenwood, D A (1987a)
"Underpinning by grouting"
Ground Engineering, Vol 20, No 3, April, pp 21–32

Hewlett, P C and Hutchinson, M T (1983)
"Quantifying chemical grout performance and potential toxicity"
Proc 8th Eur Conf on Soil Mech and Found Engg, Helsinki
AA Balkema, Rotterdam, Vol 1, pp 361–366

Houlsby, A C (1985)
"Cement grouting: water minimizing practices"
In: *Issues in dam grouting*, Geotech Engg Div, Am Soc Civ Engrs

Karol, R H (1982a)
"Chemical grouts and their properties"
Proc Conf Grouting in geotechnical engineering, New Orleans
Am Soc Civ Engrs, pp 359–377

Karol, R H (1982b)
Seepage control with chemical grout
Proc Conf Grouting in geotechnical engineering, New Orleans
Am Soc Civ Engrs, pp 564–575

Littlejohn, G S (1982)
Design of cement-based grouts
Proc Conf Grouting in geotechnical engineering, New Orleans
Am Soc Civ Engrs, pp 35–48

Littlejohn, G S, Ingle, J and Dadasbilge, K (1984)
"Improvement in base resistance of large-diameter piles founded in silty sand"
Proc 8th Eur Conf Soil Mech and Found Engg, Helsinki, 1983
AA Balkema, Rotterdam, Vol 1, pp 153–156

Mitchell, J K (1982)
"Soil improvement – state of the art"
Proc 10th Int Conf Soil Mech and Found Engg, Stockholm, 1981
AA Balkema, Rotterdam, Vol 4, pp 509–565

Mongilardi, E and Tornaghi, R (1986)
"Construction of large underground openings and use of grouts"
Proc Int Conf on Deep Foundations, Beijing, China, 1986
Deep Foundations Inst and CIGIS, Vol 2, pp 1/58–76

Skipp, B O and Hall, M J (1982)
Health and safety aspects of ground treatment material
Report 95, CIRIA, London

9.3   HYDROFRACTURE GROUTING

9.3.1   Definition

Hydrofracture grouting is the deliberate initiation and propagation of new fractures in the ground by the injection of grout under pressure in order that the fractures themselves become the channels for the grout to transfuse the ground (Figure 9.1(b)). The term also applies to the deliberate opening of existing fissures or joints. Other terms for this process are fracturing and *claquage*.

9.3.2   Principle

As the pressure of a fluid intruding into a mass of ground increases, the ground deforms away from the point of entry. When the local tensile strength of the ground material is exceeded, a fracture is initiated that will propagate away from the entry point if the pressure is maintained and new fluid introduced. In the case of existing fissures and in unconsolidated materials, the pressure causing hydraulic fracture is related to the minor principal stress in the ground before fracture; usually the vertical stress is the critical value. Hydrofracture, therefore, creates its own channels for grout entry, which are often in the form of sheets, lenses or fingers. These grout-filled channels can be used to:

- displace blocks of surrounding ground
- create impermeable barriers in the ground
- create a skeleton or tracery of more rigid material
- provide a much larger surface area of grout contact with the ground to permit subsequent permeation (see Figure 9.15).

It has to be said that all too often hydrofractures are initiated inadvertently and to no purpose as a result of bad workmanship or design.

Description and applications

Controlled hydrofracture grouting has three purposes.

1. To form impermeable boundary zones.

2. To displace blocks of the ground.

3. To create or open fractures in the ground to increase the surface area exposed to a permeation grout.

In the first case, hydrofracture grouting, with typically a thick cement-bentonite grout, is often used to form the upper boundary for a permeation treatment, particularly at the base of a surface fill. In these circumstances the use tends towards compaction grouting as subsequent injections are made.

The orientation of the fractures that are created depends on the fabric of the soil and on the *in-situ* stress conditions. Raabe and Esters (1990) suggest how the orientation of hydrofractures can change with repeated injections (see Figure 9.13) as the stress conditions change locally. Critically important are a suitable grout with an appropriate gel time, control of both injection rate and volume and records of the applied pressure.

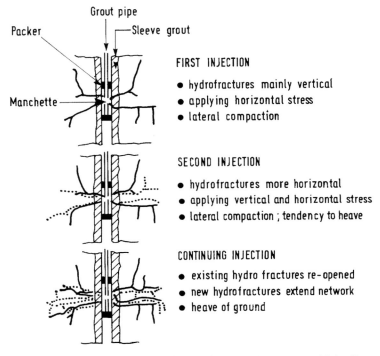

Figure 9.13   *Development of hydrofractures by repeated injection (after Raabe and Esters, 1990)*

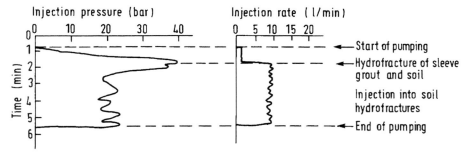

Figure 9.14   *Record of a hydrofracture injection (after Raabe and Esters, 1990)*

Figure 9.14 shows the record of a single injection at a depth of 15 m; note the high injection pressure and its build-up to the point of hydrofracture. The aim is to create a network of grout-filled hydrofractures throughout a block of ground.

In a similar way it is possible to create hydrofractures as a way of opening up the ground to receive a permeating resin grout (Figure 9.15). In this case, extremely precise control of injection rate, volume and pressure are needed.

Ayres (1985 and 1994) describes injection hydrofracture used to stabilise railway embankment slips in cohesive soils. After locating the slip surface a grid of injection points is established and the grouting proceeds upslope.

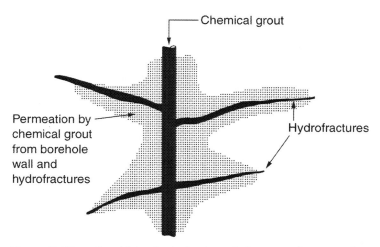

Figure 9.15   *Combining hydrofracture (to increase the area of possible grout entry) with permeation (after Caron, 1982)*

9.3.4   Limitations

The two main difficulties are the uncertainty (over the position and orientation of the hydrofractures and how much is in them) and the lack of control over the extent of the hydrofracture. Even with quite viscous grouts and the pressure diminution that occurs with distance from the injection point, the fractures can extend considerable distances. Untoward hydrofracturing has undoubtedly caused considerable wastage of grout and its appearance far from the intended zone of treatment.

9.3.5   Design

Successful use of controlled hydrofracturing requires considerable experience and a careful observational approach at each site.

9.3.6   References

Ayres, D J (1985)
"Stabilization of slips in cohesive soil by grouting"
In: *Failures in earthworks*, Proc Symp, Inst Civ Engrs, pp 424–427

Ayres, D J (1994)
"Hydrofracture grouting of landslips in cohesive soils"
In: *Grouting in the ground* (A L Bell, ed), Proc Int Conf at Instn Civ Engrs, Nov 1992
Thomas Telford, London, pp 261–272

Caron, C (1982)
"The state of grouting in the 1980s – background talk"
Proc Conf Grouting in geotechnical engineering, New Orleans
Am Soc Civ Engrs, pp 346–358

Raabe, E W and Esters, K (1990)
"Soil fracturing techniques for terminating settlements and restoring levels of buildings and structures"
Ground Engineering, Vol 23, No 4, May, pp 33–45

9.4 JET GROUTING

9.4.1 Definition

Jet grouting is the construction of hard, impervious columns in the ground by the enlargement of a drillhole using rotating fluid jets to liquefy and mix grout with, or to excavate and replace, soils (Figures 9.1 (c) and (d)). Jetting and grouting are carried out during controlled withdrawal and rotation of the drill string and the jetting head (or monitor) from the hole. There are several variations: depending on the nature and pressure of the jetting and grouting the *in-situ* soil may be mixed with the grout, partly mixed and partly removed, or wholly replaced.

9.4.2 Principle

The methods of jet grouting improve the ground by either:

a. mixing the grout into the ground material as it is disturbed by the energy of the rotating high-pressure jet of grout or of grout shrouded by air

   or

b. replacing by tremie-pumping grout into the slurry-filled cavity created by eroding the ground with rotating high-pressure water and air jets

   or

c. combining replacement and mixing using an air-shrouded water jet for eroding the soil and tremie-pumping the grout.

9.4.3 Description

Early work on the jet grouting method was carried out in the UK and was reported by Nicholson (1963). Development continued separately in Japan in the late 1960s, with the first work being described by Miki (1973) and Yahiro and Yoshida (1973). The method was applied principally to silty sands, using neat cement grouts. Several variations of the basic method then developed (see Miki and Nakanishi, 1984). There are two basic systems of construction.

1   A simple jet of grout is used to erode the ground and form either a grout column or a mixed-in-place soil-cement (the CCP method).

2   Either an air jet (the JSG system) or combined air and water jets (CJG and SSS-MAN systems) are used to erode the ground before grouting. These systems generally replace the soil rather than mixing it in place.

Jet grouting is carried out in ground principally below the groundwater table. The value of shrouding the water jet with an air jet is that in a water-filled hole the pressure of the water jet is not attenuated as quickly as an unshrouded one. The method used in UK is shown in Figure 9.16, based on the case history reported by Riddell (1986). The triple fluid monitor is shown on Figure 9.17.

Water and grouting pressures range between 20 MN/m^2 and 60 MN/m^2. Grout strengths can vary between 2 MN/mm^2 and 15 N/mm^2, with fly-ash being added to improve durability and to control excessive bleeding. The volume and quality of the grouted mass is determined by the withdrawal and rotation rates of the grout monitor, as well as by the soil type, grout mix, nozzle type and pressures used (Munfakh *et al*, 1987).

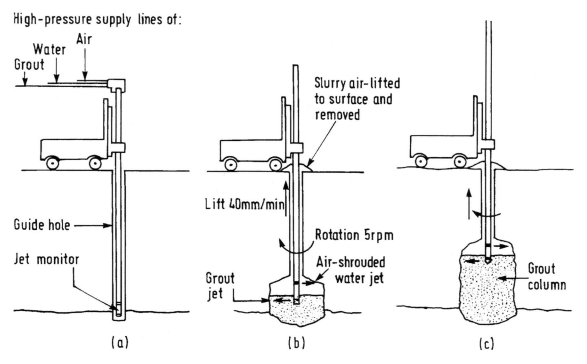

Figure 9.16   *Jet grout construction: (a) monitor lowered in guide hole; (b) jetting and grouting; (c) monitor raised as column is formed (after Riddell, 1986)*

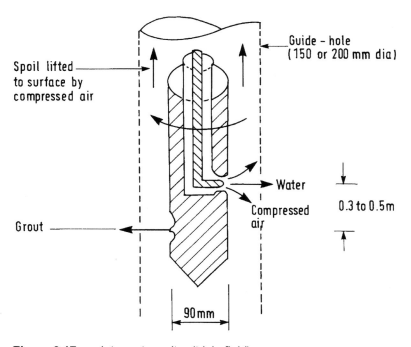

Figure 9.17   *Jet grout monitor (triple fluid)*

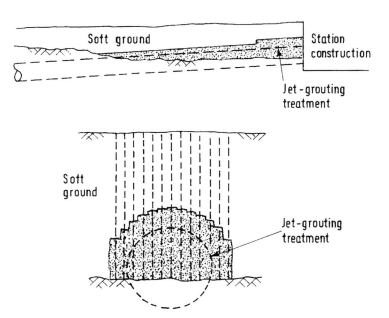

Figure 9.18 *Jet grouting for tunnel excavation in Singapore (after Lunardi et al, 1986)*

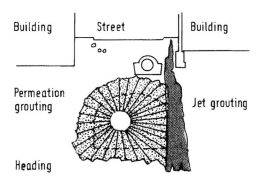

Figure 9.19 *Jet grouting to protect a building in conjunction with permeation grouting prior to tunnel construction (after Mongilardi and Tornaghi, 1986)*

These authors examine the effects of jet pressures and withdrawal rates on the grouted volume for 15 case histories in ground types ranging from clay and silt to soft rock. While similar grouted volumes can be obtained, a much slower withdrawal rate is needed for the harder, more cohesive, types of ground.

As preliminary guidance, Welsh *et al* (1986) suggest the following to relate soil type to potential grout volumes.

1. For any soil, the grout or water pressure and probe withdrawal rate are the most significant factors regarding the column of jet-grouted soil.

2. At a given grout pressure and withdrawal rate, jet-grouted volume decreases with increasing clay content.

3. As clay content increases, grout pressure must be increased and/or withdrawal rate decreased for a given jet-grouted volume.

4. It is difficult to obtain jet grout column diameters in excess of 1.5 m in stiff to hard clays using typical grout or water pressures.

5. Jet-grouted volume is not significantly affected by grain size distribution if the uniformity coefficient, (D_{60}/D_{10}), is equal to or greater than 8.

6. If the gravel-size and larger particle content of the soils is above 50 per cent, grout penetration will be reduced and be more irregular because of the tendency of the larger particles to reflect (deflect) the jet stream The result is incomplete treatment.

The density of the ground affects the sizes of the grouted columns. Typically, they could be 0.75 m in diameter in firm clays, and up to about 3 m in silty sands; 2 m is more common.

Covil and Skinner (1994) review jet grouting parameters and performance. Quality control of jet grouting is not fully standardised. The Japanese determine the size of the excavated column using ultrasonic techniques. Air/water and grout flow can be monitored continuously, but this is not always done. Water pressure and rate of lift of the monitor, together with rotation, are the most important parameters to measure. In many instances, a full-scale trial with exposure of the columns appears to be a prudent first stage in the use of jet grouting. Bertero *et al* (1988) also discuss the usefulness of ways of checking the treatment.

9.4.3   Applications

Temporary works

Most uses of jet grouting in the UK, Italy and Japan have been as a temporary measure to control groundwater, particularly in tunnels, shaft bases, and cofferdams. Examples are given by Baschieri *et al* (1983), Coomber (1985), Coomber and Bell (1985) and Mongilardi and Tornaghi (1986).

Figures 9.18 and 9.19 show two temporary uses of jet grouting – for the construction of a tunnel in soft marine clays beneath roads and buildings, and to provide temporary support to a building before excavating a tunnel. Jet grouting in the latter case was used in conjunction with chemical grouting.

Permanent works

About 25 per cent of jet grouting contracts are for permanent uses. The method can be used for many purposes, eg underpinning, permanent support, for groundwater cut-offs and to prevent scour at bridge piers.

Riddell (1986) describes the underpinning and strengthening of a length of brick tunnel in Glasgow. Figure 9.20 is a cross-section of the completed works.

Munfakh *et al* (1987) list 18 jet grouting usages, both permanent and temporary. Many applications are described in the proceedings of the conference, *Grouting in the ground* (Thomas Telford, 1994) including the major stabilisation work for the Kingston Bridge, Glasgow (Carruthers *et al*, 1994 and Coutts *et al*, 1994) and the formation of a jet grouting curtain below a dam in Kenya (Harris and Morey, 1994).

A novel use of the basic method is for clean-up of soil contaminated by phenols (*Ground Engineering*, 1991). The jetting, by water only, was used to erode the soil and enable it to be air-lifted to a soil washing unit. A large proportion of the cleaned soil is mixed with cement and aggregate and tremied back into the column. This modification of the technique is essentially the same as the replacement method of jet grouting although perhaps not strictly a grouting process: it is, however, a ground improvement technique.

9.4.4   Limitations

Jet grouting is not applicable to fibrous peats, but it might be successful in amorphous peat. The possible effect on the grout of acidic groundwater in peaty or contaminated conditions should be considered in the mix design. In made or filled ground, the grout column cannot be defined with certainty. Glacial tills might be difficult to treat because the jets might flush out the granular pockets leaving the clay matrix largely intact.

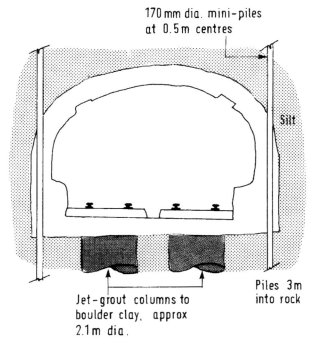

170 mm dia. mini-piles
at 0.5 m centres

Silt

Jet-grout columns to
boulder clay, approx
2.1 m dia.

Piles 3m
into rock

Figure 9.20 *Jet grouting to underpin the invert of a railway tunnel (after Riddell, 1986)*

Removal of the slurry or spoil at the surface can also be a problem, as it can be about two to three times the grouted volume. This slurry is partly grout. The situation can be eased by using the mix-in-place method rather than complete replacement. Effective removal, *via* the borehole around the monitor, is essential to minimise ground heave, particularly in very soft clays. Upward surface movements of up to about 0.5 m, or more were recorded in Singapore. Ammonia was also released into tunnels when the alkaline cement at high temperatures interacted with the organic contents of the soft clays. Heave effects can be minimised by using pre-drilled holes through soft ground.

For long-term or permanent usages, the durability of the grout mix has to be considered.

9.4.5   Design

The technique is still innovative and, while the body of experience is increasing, design is still empirical. Guidance is given by Covil and Skinner (1994) and Bell and Burke (1994) and other papers to the *Grouting in the ground* conference.

9.4.6   References

Baschieri, F, Jamiolkowski, M and Tornaghi, R (1984)
"Case history of a cut-off wall executed by jet grouting"
Proc 8th Eur Conf Soil Mech and Found Engg, Helsinki, 1983
AA Balkema, Rotterdam, Vol 1, pp 121–126

Bell, A L (ed) (1994)
Grouting in the ground, Proc Int Conf at Instn Civ Engrs, November 1992
Thomas Telford, London

Bell, A L and Burke, J (1994)
"The compressive strength of ground treated using triple system jet grouting"
In: *Grouting in the ground* (A L Bell, ed), Proc Int Conf at Instn Civ Engrs, Nov 1992
Thomas Telford, London, pp 525–538

Bertero, M, Marchi, G, Merli, M and Paviani, A (1988)
"Foundation improvement by jet grouting of a historical building in Cervia, Italy"
Proc Int Symp on Engineering geology of ancient works, monuments and historical sites
(Marinos and Konkis, eds)
AA Balkema, Rotterdam, Vol 1, pp 381–391

Carruthers, D, Coutts, D, McGown, A and Greenwood, D (1994)
"Background to the design of the quay wall stabilisation works at Kingston Bridge, Glasgow"
In: *Grouting in the ground* (A L Bell, ed), Proc Int Conf at Instn Civ Engrs, Nov 1992
Thomas Telford, London, pp 417–432

Coomber, D B (1985)
Tunnelling and soil stabilization by jet grouting
Tunnelling '85, Instn Mining and Metallurgy, London, pp 277–283

Coomber, D B and Bell, A L (1985)
"Groundwater control by jet grouting"
Proc 21st Reg Mtg of Engg Grp of Geol Soc, Sheffield, 1985

Covil, C S and Skinner, A E (1994)
"Jet grouting – a review of some of the operating parameters that form the basis of the jet grouting process"
In: *Grouting in the ground* (A L Bell, ed), Proc Int Conf at Instn Civ Engrs, Nov 1992
Thomas Telford, London, pp 605–630

Coutts, D, Hutchinson, D E and Essler, R D (1994)
"Specification, planning and construction of quay wall stabilisation works at Kingston Bridge, Glasgow"
In: *Grouting in the ground* (A L Bell, ed), Proc Int Conf at Instn Civ Engrs, Nov 1992
Thomas Telford, London, pp 433–454

Ground Engineering (1991)
"Carbolic clean-up"
Ground Engineering, Vol 24, No 4, May, pp 14–15

Harris, R R W and Morey, J (1994)
"Construction of a jet mix cut-off for Thika dam, Kenya"
In: *Grouting in the ground* (A L Bell, ed), Proc Int Conf at Instn Civ Engrs, Nov 1992
Thomas Telford, London, pp 487–498

Lunardi, P, Mongilardi, E and Tornaghi, R (1986)
"Il preconsolidamento mediante jet grouting nella realizzazione di opere sotteraneo"
Proc Int Cong. Large Underground Openings, Florence, 1986, Vol 2, pp 601–612

Miki, G (1973)
"Chemical stabilisation of sandy soils by grouting in Japan"
Proc 8th Int Conf Soil Mech and Found Engg, Moscow, p 395

Miki, G and Nakanishi, W (1984)
"Technical progress of jet grouting method and its newest type"
Proc Int Conf on in-situ soil and rock reinforcement, Paris, 1984, Vol 1, pp 195–200

Mongilardi, E and Tornaghi, R (1986)
"Construction or large underground opening and use of grout"
Proc Int Conf on Deep Foundations, Beijing, 1986
Deep Foundation Inst and CIGIS, Vol 2, pp 1/58–76

Munfakh, G A, Abramson, L W, Barksdale, R D and Juran, I (1987)
"*In-situ* ground reinforcement"
In: *Soil improvement – a ten year update* (J P Welsh, ed)
Geotech Special Publication 12, Am Soc Civ Engrs

Nicholson, A J (1963)
Discussion in: *Grouts and drilling sands in engineering practice*
Proc Conf Instn Civ Engrs, Butterworths, London, pp 108–109

Riddell, J B (1986)
"Strengthening works in High Street tunnel, Glasgow"
Proc Instn Civ Engrs, Vol 80, Pt 1, pp 1233–1249

Welsh, J P, Rubright, R M and Coomber, D B (1986)
"Jet grouting for support of structures"
Spring convention, Seattle, 1986, Am Soc Civ Engrs

Yahiro, T and Yoshida, H (1973)
"Induction grouting method utilising high speed water jet"
Proc 8th Int Conf Soil Mech and Found Engg, Moscow, 1973, pp 359–362 and 402–403

9.5 COMPACTION, SQUEEZE AND COMPENSATION GROUTING

9.5.1 Definitions

1. *Compaction grouting*, also called *displacement grouting*, is the controlled injection, without causing hydrofracture, of a mortar as an expanding bulb to displace, and hence compact, loose soils (Figures 9.1(e) and 9.21). (Note: this is a more restricted definition than often used elsewhere. Often the term "compaction grouting" is used to include squeeze and compensation grouting and deliberate hydrofracture).

2. *Squeeze grouting* is the controlled high-pressure injection of stiff grout into broken ground, usually rock at considerable depth, to displace the blocks and strengthen the ground mass (Figure 9.1(f)).

3. *Compensation grouting* is the responsive and timely use of compaction or hydrofracture grouting as an intervention between an existing structure and an engineering operation (particularly tunnel excavation) to counteract any movement of the ground that would otherwise affect the existing structure. Ideally, the grout should be injected before the potential wave of settlement has progressed from the underground excavation to the foundation to be protected. In practice, the grouting is often done in advance of the excavation or after or both. There are perhaps two reasons for this. First, the settlement wave occurs rapidly, ie in hours, so there is little time for what is a complex grouting operation; second, there is concern about the injection pressures affecting the stability of the advancing excavation face. Much compensation grouting appears therefore to be jacking either before or after settlement of a building.

9.5.2 Principle

Compaction, squeeze and compensation grouting displace the ground mass or blocks of ground by the insertion of a body of stiff grout. These methods represent a range of processes from intrusion, controlled so as to prevent hydrofracture, to the deliberate creation of hydrofractures. Operational control of these types of grouting is based upon thorough, concurrent measurements of ground and structure movements.

9.5.3 Descriptions and applications

Compaction grouting

Compaction grouting is used in soils, typically those finer than medium sand. The mortar is injected at the base of a previously drilled borehole. Careful control of the pressure build-up is needed to prevent hydrofracture and to create a substantially spherical or cylindrical bulb as the grout mass expands. The effect of the pressure from the injected mortar as it is transferred to the surrounding ground is to compress its weaker zones. The uplift forces that are generated can be used to relevel ground slabs, road pavements or buildings. Alternatively, a series of grout bulbs can be intruded into the ground below a structure as a method of underpinning. The grout has typical compressive strengths, when set, of about 3 N/mm^2 and is usually a mix of sand and cement, but pozzolans or lime are also used.

Stiller (1982) reports the use of compaction grouting to lift and stabilise the foundations of two masonry pump-house buildings, founded on clayey sand fill.

After distress was first noticed, 80 mm of differential settlement was measured. Compaction grouting under the spread footings was taken down 5–10 m to the underlying sandstone. The grouting process raised the buildings by about 80 mm. Other examples of lifting and levelling buildings are given by King and Bindhoff (1982) and by Greenwood (1984 and 1987). Sealy and Bandimere (1987) describe the underpinning of houses on soft fills and silts with boulders, carried out in harsh winter conditions. Greenwood (1984) gives details of the underpinning of a gas-holder in Rhyl.

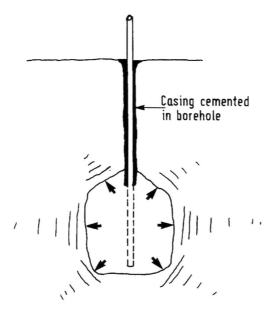

Casing cemented
in borehole

Figure 9.21   *Principle of compaction grouting (after Warner, 1982)*

Donovan (1984) reports the use of compaction grouting to improve a layer of loose sand beneath a hospital in a seismically active area. Henry (1987) reports the use of compaction grouting to deal with solution features. In one case, compaction grouting was used together with dynamic compaction for the foundations to a power station. In another, grouting was used to stabilise a sinkhole beneath a house. Welsh (1988) gives details of two further case histories where compaction grouting improved ground over karstic limestone. Sinkholes could then not develop. Francescon and Twine (1994) also treated solution features in the chalk at Norwich by compaction grouting.

Techniques essentially the same as for compaction grouting are also used to raise slabs, ie "slabjacking". Often a cavity to receive the grout is first created under the slab. Polatty (1982) gives details of the repair of road slabs. Grout was designed both to relevel the slab and to be permeable enough to permit drainage. Younger and Rananand (1985) give details of the raising of the national highway between Bangkok and Nakhon Pathon, laid directly on a lateritic sub-base.

Bruce and Joyce (1983) used cement/bentonite to raise the concrete slabs forming part of the auxiliary spillway to Tarbela Dam The underlying rock is karstic limestone. Uplifts of between 0 mm and 127 mm were achieved.

Squeeze grouting

Squeeze grouting is used to consolidate fractured rock zones in deep mines and tunnels and, provided there is careful control of pressures it can be used in fractured ground at shallower depths (Greenwood and Hutchinson, 1982). Its use for the restoration of total soil stresses in the ground (London Clay) above a collapsed tunnel at Canterbury is described by Vaughan *et al* (1983) and by Greenwood (1987). The grout is injected into existing fractures, rather than creating new ones by hydrofracture. In certain circumstances, hydrofracturing can be used for the same purpose, eg in the puddle clay core of an earth dam, where the core may have cracked internally because of arching and differential settlement between the core and the shoulders of the dam

Seven examples of tunnels treated by squeeze grouting are given by Greenwood and Hutchinson (1982). Its use for the ground above the collapsed tunnel in Canterbury is illustrated in Figure 9.22. Grout injection under gravity was used first to fill voids in the collapsed stiff-clay rubble in and above the tunnel prior to the squeeze grouting.

Compensation grouting

Compensation grouting (Figure 9.23) is used to prevent structures being affected by the ground displacements, particularly settlement, that would result from nearby tunnelling or other deep excavation (Linney and Essler, 1994). The injected body of ground can be considered as compensating for the ground loss during and subsequent to the excavation in relation to the structure being protected. The grouting may combine the methods of compaction and, if controllable, hydrofracture grouting, not as underpinning or to restore the level of a building, but to maintain the structure at its existing level. Thus the operation has to be responsive both to the ongoing excavation works and to instantaneous observations on the structure to be protected.

The technique is still developing as its viability improves with electronic control and measurement systems. Early uses are described by Warner (1982) and a trial prior to tunnel construction is reported by Linney and Essler (1994). While full case histories have yet to be published, the extensive use of compensation grouting on the Jubilee Line Extension Project proved successful, particularly in the area of prestigious buildings of Westminster and St James's.

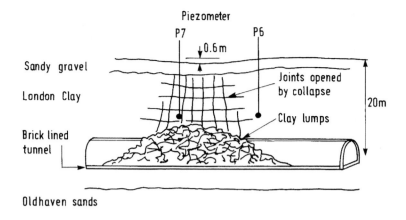

Oldhaven sands

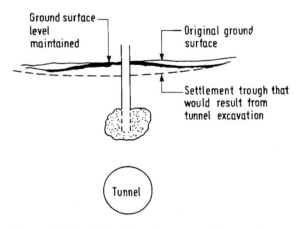

Figure 9.22 *Squeeze grouting for restoration of total soil stress above collapsed tunnel (after Greenwood, 1987b)*

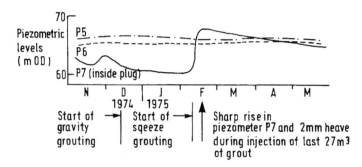

Figure 9.23 *Principle of compensation grouting*

9.5.4   Limitations

Compaction and compensation grouting are limited to soils, usually finer than a medium sand. Squeeze grouting is used for rubbly ground but at sufficient depth for there to be a substantial pressure to work against. All three require careful management and recording of the injection rate and pressure whether to avoid hydrofracture or, insofar as it is possible, to fracture deliberately. Should pressures build up too quickly, sudden losses will occur with break-out as the grout finds the weakest zones of the ground. Very close observation of the behaviour of the ground is needed, eg to avoid surface heave and, in the case of compensation grouting, on ground/structure movement monitoring systems linked directly to the grouting control (see Price and Wardle, 1994).

9.5.5 Design

Perhaps the most comprehensive guidance is given by Warner (1982) as he surveyed 30 years of compaction grouting practice. Squeeze grouting is addressed by Greenwood and Hutchinson (1982). For a discussion of compensation grouting, see Linney and Essler (1994) and Mair's report (Mair, 1994) at the 1992 conference *Grouting in the ground* (Bell, 1994).

9.5.6 References

Bell, A L (ed) (1994)
Grouting in the ground, Proc Int Conf, Instn Civ Engrs, November 1992
Thomas Telford, London

Bruce, D A and Joyce, G M (1983)
"Slabjacking at Tarbela dam, Pakistan"
Ground Engineering, Vol 16, No 3, pp 35–39

Donovan, N C (1984)
"Soil improvement in a sensitive environment"
Proc Conf in situ Soil and Rock Reinforcement
Presses de L'Ecole Nationale des Ponts et Chaussées, Paris, pp 185–190

Francescon, M and Twine, D (1994)
"Treatment of solution features in Upper Chalk by compaction grouting"
In: *Grouting in the ground* (A L Bell, ed), Proc Int Conf, Instn Civ Engrs, Nov 1992
Thomas Telford, London, pp 327–348

Greenwood, D A (1984)
"Re-levelling a gas holder at Rhyl"
Q J Engg Geol, Vol 17, pp 319–326

Greenwood, D A (1987)
"Underpinning by grouting"
Ground Engineering, Vol 20, No 3, April, pp 21–32

Greenwood, D A and Hutchinson, M T (1982)
"Squeeze grouting unstable ground in deep tunnels"
Proc Conf on grouting in geotechnical engineering, New Orleans
Am Soc Civ Engrs, pp 631–651

Henry, J F (1987)
"The application of compaction grouting to Karstic foundation problems"
Proc 2nd Multidisciplinary Conf on Sinkholes, Orlando, Florida, 1987, pp 447–450

King, J C and Bindhoff, E W (1982)
"Lifting and leveling heavy concrete structures"
Proc Conf Grouting in geotechnical engineering, New Orleans
Am Soc Civ Engrs, pp 722–737

Linney, L F and Essler, R D (1994)
"Compensation grouting trial works at Redcross Way, London"
In: *Grouting in the ground* (A L Bell, ed), Proc Int Conf, Instn Civ Engrs, Nov 1992
Thomas Telford, London, pp 313–326

Mair, R J (1994)
"Report Session 4, Displacement"
In: *Grouting in the ground* (A L Bell, ed), Proc Int Conf at Instn Civ Engrs, Nov 1992
Thomas Telford, London, pp 375–384

Polatty, J M (1982)
"Highway grouting – a permeable grout provides drainage below a concrete pavement"
Proc Conf Grouting in geotechnical engineering, New Orleans
Am Soc Civ Engrs, pp 849–858

Price, G and Wardle, I (1994)
"The use of computer controlled monitoring systems to control compensation and fracture grouting"
In: *Grouting in the ground* (A L Bell, ed), Proc Int Conf at Instn Civ Engrs, Nov 1992
Thomas Telford, London, pp 203–214

Sealy, C O and Bandimere, S W (1987)
"Grouting in difficult soil and weather conditions"
J Perf of Constrn. Facilities, Am Soc Civ Engrs, Vol 1, No 2, May, pp 84–94

Stiller, A N (1982)
"Compaction grouting for foundation stabilisation"
Proc Conf Grouting in geotechnical engineering, New Orleans
Am Soc Civ Engrs, pp 923–937

Vaughan, P R, Kennard, R M and Greenwood, D A (1983)
"Squeeze grouting of stiff fissured clay after a tunnel collapse"
Proc 8th Eur Conf Soil Mech and Found Engg, Helsinki, 1983
AA Balkema, Rotterdam, Vol 1, pp 171–177

Warner, J F (1982)
"Compaction grouting – the first thirty years"
Proc Conf on Grouting in geotechnical engineering, New Orleans
Am Soc Civ Engrs, pp 694–707

Welsh, J P (1988)
"Sinkhole rectification by compaction grouting"
Proc Conf Geotechnical aspects of Karst terrains, May 1988
Am Soc Civ Engrs, pp 115–132

Younger, J S and Rananand, N (1985)
"Ground improvement works: use of geotextiles and modern piling methods in Thailand"
Proc Symp on Recent developments in ground improvement techniques, Bangkok, 1982
AA Balkema, Rotterdam, pp 255–276

9.6 CAVITY FILLING

9.6.1 Definition

Cavity infilling is the filling of man-made or natural cavities. At one extreme it includes the filling of abandoned mine workings or the solution features of karstic limestones; at the other it includes the annulus grouting around tunnel linings (Figure 9.1(g)).

9.6.2 Principle

Grout is injected at low pressures or simply by gravity to the base of the cavity to displace any water and air within the cavity to fill it as completely as possible or to create a stable support below the roof of the cavity or between the surrounding ground and a structure.

9.6.3 Description and applications

In terms of quantities and numbers of projects, the filling of abandoned mine workings and solution features represents the largest use of grout.

Abandoned mine workings

Mansur and Skouby (1970) give details of the use of cement grouting to fill old workings beneath a building under construction. Grout procedures are also reported by Littlejohn (1979) and Healy and Head (1984). Figure 9.24 shows a typical sequence of operations. The use of grouting for old shafts is also described by Healy and Head (1984). Ward and Hills (1987) describe trials for the use of colliery spoil in a rock-paste form for filling abandoned limestone mines, work described by Cole and Stevens (1987).

Grouting karstic rock

Ortiz-Suarez and Agrelot (1982) describe two case histories where cement grout was pumped through a series of 150 mm-diameter holes into relatively open caverns, or very loose soils within caverns. Grouting was carried out until uplift and surface cracking occurred around the grouting points. Davies and Lord (1984) report the grouting of cavities beneath the foundations of an 18-storey building in Al-Khobar, Saudi Arabia. Cement-bentonite was used with perimeter borings at 2.5 m centres and seven borings beneath each footing.

Sinkholes can also be treated by grouting. Several case histories are described by Garlanger (1984) and Ryan (1984). Raghu et al (1987) give details of the use of a percussive probe to judge the effectiveness of grouting cavernous ground.

Where the cavities are largely or wholly infilled different techniques are needed. Golder (1982) gives brief details of the foundations of a 45 m-diameter reactor building applying a bearing pressure of 380 kN/m². The building was underlain by about 20 m of highly karstic limestone. Cement grouting was used initially at 2.5 m then at 1.2 m centres to compact loose sand within cavities. The sand was grouted chemically to minimise the possibility of further solution.

Zhang and Huo (1982) and Zhang (1988) used high-pressure grouting in karstic caves up to 100 m high, and infilled with soft clay. Drillholes into the caves were at 1.5–2.5 m centres. The caves were injected with cement grout using the descending stage method (Figure 9.6b) at pressures of up to 60 bar. The principal mechanism of improvement was from the hydrofracturing, reinforcing the soft clay infilling. Extrusion of the soft clay also occurred together with consolidation and chemical hardening.

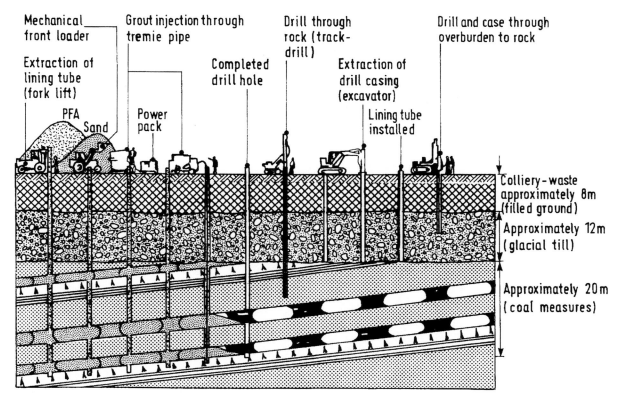

Figure 9.24 *Sequence of operations adopted at Merthyr Tydfil, Wales (after Littlejohn (1979))*

9.6.4 Limitations

The main difficulties with treating cavities in the ground is obtaining sufficient knowledge about their position, shape and infilling. These will determine the methods of grouting that would be appropriate. For large-scale works, the aim is usually to choose a low-cost grout or filling material.

9.6.5 Design

Guidance is given by Healy and Head (1984) on the treatment of abandoned mine workings. In the use grouts for infilling solution features, the specific site circumstances vary so much that each treatment (and often a combination of treatment techniques) will be designed specially.

9.6.6 References

Cole, K W and Stevens, D W (1987)
"Infilling abandoned limestone mines in the Black Country"
Proc 23rd Ann Conf of Engg Grp of Geol Soc (F G Bell, M G Culshaw, and J C Cripps, eds)

Davies, J A G and Lord, J A (1984)
"The effects of cavities in limestone on the construction of a high-rise building in Al-Khobar"
Proc Symp on Geotechnical problems in Saudi Arabia, University of Riyadh, May 1981

Garlanger, J E (1984)
"Remedial measures associated with sinkhole-related foundation fishers"
In: *Sinkholes: Their geology, engineering and environmental impact* (B F Beck, ed)
AA Balkema, Rotterdam, pp 413–418

Golder, H Q (1982)
"On grouting, discussion, in Session 12 Soil improvement"
Proc 10th Int Conf Soil Mech and Found Engg, Stockholm, 1981
AA Balkema, Rotterdam, Vol 4, pp 954–955

Healy, P R and Head, J M (1984)
Construction over abandoned mine workings
Special Publication 32, CIRIA, London

Littlejohn, G S (1979)
"Surface stability in areas underlain by old coal workings"
Ground Engineering, Vol 12, No 3, March, pp 22–30

Mansur, C I and Skouby, M C (1970)
"Mine grouting to control building settlement"
J Soil Mech and Found Div, Am Soc Civ Engrs, Vol 96, SM2, March, pp 511–522

Ortiz-Suarez, C A and Agrelot, J C (1982)
"Grouting caverns and soft zones by concrete pumps"
Proc Conf Grouting in geotechnical engineering, New Orleans
Am Soc Civ Engrs, pp 987–999

Raghu, D, Antes, D R and Lifrieri, J J (1987)
"Use of percussion probes to determine rock mass quality of cavernous carbonate formations before and after grouting"
In: *Karst hydrogeology: engineering and environmental applications* (B F Beck and W L Wilson, eds)
AA Balkema, Rotterdam, pp 389–396

Ryan, C R (1984)
"High volume grouting to control sinkhole subsidence"
In: *Sinkholes: their geology, engineering and environmental impact* (B F Beck, ed)
AA Balkema, Rotterdam, pp 413–418

Ward, W H and Hills, D L (1987)
"Rock paste for filling abandoned mines"
Ground Engineering, Vol 20, No 5, July, pp 29–33

Zhang, Z M (1988)
"Application of hydrofracture principle to grouting in deep foundations"
Proc Int Conf on Deep Found, Beijing, September
China Building Industry Press, Beijing, Vol 1, pp 1/10–15

Zhang, Z M and Huo, P S (1982)
"Grouting of the karstic caves with clay fillings"
In: *Grouting in geotechnical engineering* (W H Baker, ed)
Am Soc Civ Engrs

10   Improvement by thermal stabilisation

Even in the temperate UK, everyone is familiar with the way that surface soils are hardened, albeit temporarily, by frost and hot, dry weather. The removal of heat from the soil turning its pore water into ice is a very powerful technique rendering the ground impermeable and, for unconsolidated materials, making them stronger. Applying heat to clays to drive out free pore water and, at higher temperatures, the water adsorbed on particle surfaces, creates a very hard, durable material – in effect, the same methods as when making brick or mud (adobe) building blocks.

Ground freezing is a long established and particularly effective method of ground stabilisation for temporary works. Ground heating is rare, but when it has been used its purpose was longer-term improvement.

10.1   ARTIFICIAL GROUND FREEZING

10.1.1   Definition

Artificial ground freezing is where refrigeration of the ground converts pore water to ice, binding together the soil particles as a form of improvement.

10.1.2   Principle

While all types of soil and rock containing moisture can be frozen, there has to be sufficient pore or joint water to achieve impermeability and increase in strength. Frozen conditions are created by circulating a cold medium through a series of freeze tubes positioned close enough together to form an ice wall, cofferdam or barrier. This wall will be virtually impermeable and the soil/ice structure will have greatly enhanced strength. Such properties are sought where linear structures have to be installed through very mixed soils with a high groundwater table. The effect is to produce a strong impermeable barrier so that work can proceed in dry conditions. Freezing is usually a temporary measure, with the groundwater regime restored after the thawing of the ice wall or barrier. It can be used permanently, however, in Arctic regions to maintain permafrost conditions beneath heated structures.

10.1.3   Description

The history of freezing is described by Harding and Glossop (1939), with the first use in South Wales in 1862 to sink a shaft. The Poetsch process was then developed after 1880 and in the 1930s was used on such contracts as the Antwerp tunnels, the Moscow Subway, the Grand Coulee dam and the 300 m-deep Moudsworth shaft, Cheshire. There are two basic systems of refrigeration (see Table 10.1).

The first is the "closed" or mechanical, system, which circulates ammonia or freon to chill calcium chloride brine, the secondary refrigerant and the medium of heat transfer, which is circulated through the freeze-tube circuit. The brine is usually at a temperature between -25° and -30°C, but can operate to -60°C (Gallavresi, 1985). Figure 10.1 shows the scheme of a typical arrangement. Heat extracted from the groundwater is then dissipated to the atmosphere via cooling towers or evaporative condensers. It usually takes three to five weeks to form the ice wall or barrier.

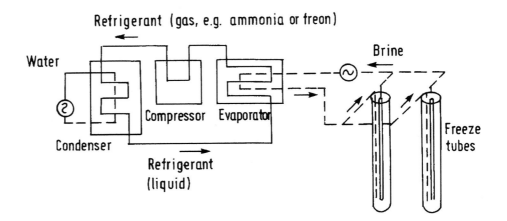

Figure 10.1 *Closed system of ground freezing*

Table 10.1 *Open and closed systems of artificial ground freezing (after Harris and Pollard, 1985)*

	Closed system	Open system
Refrigerant	Ammonia or freon (primary) calcium/chloride brine (secondary)	Liquid nitrogen
System	circulating	nitrogen gas expendable
Plant	compressor, condenser, evaporator, pumps	Storage tank, pumps (liquid nitrogen supplies delivered to site)
Services	electricity, water	—
Noise	of plant	negligible
Typical freeze period	three to 12 weeks	three to seven days
Temperatures of freeze tubes	-25°C to -35°C	-100°C to -196°C

The second, "open" or cryogenic, system uses liquid nitrogen as an expendable refrigerant. Intense refrigeration occurs when the liquid is caused to vaporise in the freeze tubes. The resultant gases are exhausted to the atmosphere. Figure 10.2 shows the scheme of an open system. Treatment of the ground is very rapid compared to the closed system, with the liquid nitrogen operating at -196°C.

Whenever possible, the freeze tubes are recovered after use, although sometimes they have to be grouted into the drillholes. They are usually high-quality steel tubes with threaded connections 100–150 mm in diameter; the larger-diameter pipes are required for the closed system. Internal feed pipes within freeze tubes are usually about 40–50 mm diameter.

Accurate drilling of holes for the installation of the freeze tubes is important for the success of freezing, particularly when the treatment is for a deep shaft or mine. For depths to about 250 m, Harris and Pollard (1985) suggest that it is usually possible to maintain verticality by balancing the drill string weight while maintaining sufficient weight on the bit to cut the ground efficiently. Directional drilling techniques are used below about 250 m. For horizontal freeze tubes, drilling can be carried out using the outer element of the freeze tube acting as the drill pipe.

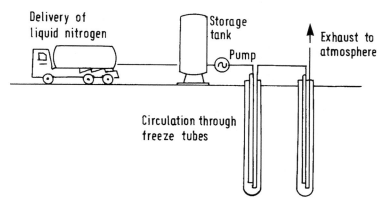

Figure 10.2   *Open system of ground freezing*

10.1.4   Applications

Artificial ground freezing is usually thought to be most frequently associated with either the sinking of deep shafts or the excavation of mining drifts, but there have been many uses for civil engineering operations at shallow depth, ie tunnels, shafts and caissons. The technique is also used on occasions to "rescue" shafts and tunnels in difficulties with bad or inadequately treated ground. These usages are temporary, although they may be in operation for several months or more. A long-term usage is the containment of liquid natural and petroleum gases.

For large-scale projects, or where the required temporary improvement is for a long time, the closed, mechanical (brine) system is normally used. The relatively simple, flexible, open cryogenic (nitrogen) system has advantages for small-scale works where speed is critical or where the period of ice maintenance is relatively short. Additionally, both systems may be used in combination. Gallavresi (1985) describes several case histories in which preliminary freezing was achieved quickly using nitrogen, to be replaced by brine pumping for continuing the freeze.

Klein (1988) gives a global overview of using freezing in the construction of shafts to diameters of 8 m and depths of 1400 m. Harvey and Martin (1988) describe the freezing operations for extending the British Coal Asfordby mine shafts from a depth of 275–465 m.

The use of artificial ground freezing in tunnelling is reviewed by Harris (1988). Rohde *et al* (1987) gives examples of the formation of frozen arches in the ground above and to the side of tunnel excavations. Tables 10.2 and 10.3 list features of these operations.

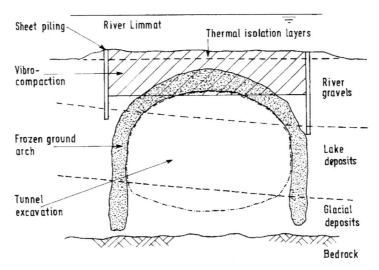

Figure 10.3 *Cross-section of River Limmat Tunnel (after Gysi and Mader, 1987)*

Table 10.2 *Ground freezing to create arches of frozen ground around tunnel excavation (after Rohde et al, 1987)*

Tunnel	Design thickness of frozen arch (m)	Overburden depth (m)	Tunnel width (m)
Metro, Helsinki	2.0–2.5	22	6.5
Hisingen, Göteborg	1.5	18	3.5
Tverrfjellet, Dovre	3.0	35	8.0
Sewer, Oslo	1.0	15	2.0
Milchbuck, Zürich	1.5–2.0	10–15	15.0
Metro, Frankfurt	1.0–1.4	6-10	7.0
Abelkrysset, Oslo	4.0	(9)	(14)
Metro, Zürich	1.4–2.0	2-4	14.0
Highway tunnel, Oslo	4.0	37.5	16.0

Table 10.3 *Performance of freezing tunnel excavation (after Rohde et al, 1987)*

Tunnel	Installed freeze capacity per metre of freeze tube (kcal/h/m)	Time for freezing (week)	Average temperature (°C)
Metro, Helsinki	58	7–12	-10
Hisingen, Göteborg	88	5	
Tverrfjellet, Dovre	87	4.3	-20
Sewer, Oslo (a)	92	6	
Sewer, Oslo (b)	143	3.5	-15
Milchbuck, Zürich	<500	1.5	-10
Metro, Frankfurt	(307)	1	-15
Abelkrysset, Oslo	50	6	-15
Metro, Zürich	80–100	2	-10
Highway tunnel, Oslo	96	9	-20

Note: see Table 10.2 for tunnel geometries.

This type of frozen-ground temporary support for tunnel excavation was used on the Limmat River tunnel, Switzerland (Gysi and Mader, 1987). Figure 10.3 shows a cross-section of the works. This case is particularly interesting because of its combination of ground improvement techniques. As well as artificial ground freezing, the contractor also used vibro-compaction to densify the river gravels, sheet piling barriers to groundwater flow, and jet grouting to extend the cut-off below the river wall.

At the 6th International Symposium on Ground Freezing held in 1991 it was reported that grouting techniques were combined with freezing to minimise water flow problems in permeable situations (Harris *et al*, 1992). The reporters show a combined application of freezing and grouting (after Jessberger, 1992) to create a frozen ring of ground around a proposed tunnel excavation. This is shown in Figure 10.4. The purpose of the grouted barriers was not to exclude but to contain specially added water to enable ground above the water table to be frozen. This addition of water to the ground is termed "irrigation".

A more straightforward scheme for a surface excavation is shown in Figure 10.5. This assumes that the ice walls form an adequate cut-off into a competent underlying station. If not, then as for the creation of underground storage tanks of liquid natural and petroleum gases, the base as well the sides would have to be frozen.

To monitor the effect and progress of the freeze operation, three variables are usually measured.

1. Temperatures at various points in the ground using thermistors or thermocouples.

2. Ground heave for near-surface works.

3. Piezometric changes in groundwater conditions.

10.1.5 Limitations

Artificial ground freezing is probably the most certain method of groundwater exclusion and can be used in all types of ground materials. Gallavresi (1985), Jessberger (1985), and Harris (1988) describe limitations on its use. As the process must have groundwater to work, then water quality, quantity and movement need to be determined. Shuster points out that water is one of the least certain elements in the design of freezing systems.

Dissolved salts or saline conditions affect the freezing point, strength and formation of the frozen ground. Clayey soils may not freeze; brine pockets can form; and excavation through natural salt deposits can be difficult. Seepages usually occur at the contact zone just above the salt. Petroleum products also prevent ground being strengthened by freezing at reasonable temperatures.

Insufficient interstitial water, and water in cavities or in very pervious rock or ground consisting of cobbles, boulders and rock debris can be difficult to freeze reliably if the water circulates too much and retards the freezing rate, because of its capacity to transfer heat.

Groundwater movement also determines the freezing process that is applicable. Flows of up to about 2 m/day can be handled using the closed (brine) system at about -20°C. If flows are greater than this then "windows" of unfrozen ground may be left. Grouting or pumping may have to be used with or during the freezing process. Continuous pumping from an excavation may cause the ice wall to deteriorate. Multiple aquifers also need careful consideration to ensure that there are no flow connections along the drillholes used to install the freeze pipes. Shafts should be covered during excavation to minimise the effects of rainfall, sunlight and convection currents of warm air.

Ground movement caused by the freezing process occurs as heave during freezing, and as consolidation during and after thawing. The expansion of the ground results from the freezing of the porewater and, more significantly, from the formation of ice lenses. The effect is more pronounced in clays and silts than in granular ground. Frost heaves of about 50 mm are not uncommon, and can be up to about 100 mm. Jessberger (1985) suggests that frost heaves can occur at a rate of 5 mm/day during freezing. Movements can be much larger for frozen organic silts and within shafts where no pressure relief has been allowed. Freezing pressures vary between 100 kN/m^2 and 2000 kN/m^2 (Jessberger, 1985).

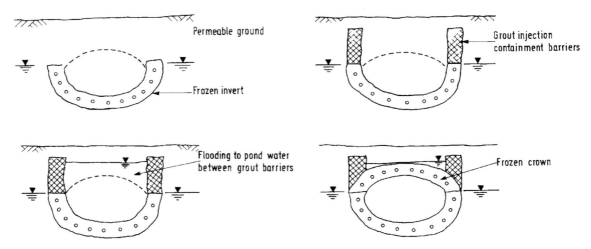

Figure 10.4 *Control of water irrigation for freezing above water table (after Harris et al, 1992)*

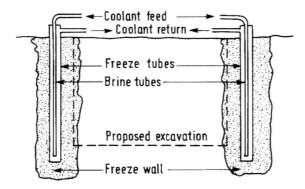

Figure 10.5 *Typical scheme of freezing for a surface excavation*

Thaw consolidation is usually about 20–25 per cent greater a movement than the frost heave. It can be controlled by the use of compaction grouting to fill the voids in the ground. Note that freezing can destroy the structure of organic soils such as peats.

As with most methods for dealing with ground and groundwater, it is prudent to install a freezing system with considerable reserve capacity over what is expected to be the necessary minimum.

Safety considerations and hazards related to ground freezing are described by Skipp and Hall (1982). In particular, loss of gaseous refrigerant, either ammonia or nitrogen, causes oxygen deficiency locally with the risk of asphyxiation. Gas monitors and masks with breathing sets should be available. Cold components, particularly metal, can cause "cold burns" if touched and severe skin damage if gripped. Leather gloves are essential, particularly with nitrogen systems.

10.1.6 Design

The design of a freezing installation requires study of two elements.

1. Heat flows (or transfers) and the thermal evaluation of the system.

2. Structural analysis of the strength of the improved ground based on stress, strain and time.

The thermal analysis is comparable to seepage or consolidation analysis. It is governed by the latent heat of fusion of water on freezing and by groundwater flow in and out of the treated zone. Kay and Perfect (1988) cover this analysis comprehensively, including the effect of solutes on the formation of ice.

Structural analysis and the deformation of frozen ground is assessed in detail by Sayles (1988), who describes various models for examining creep behaviour, although Harris (1988) points out that there is no generally agreed relationship as yet. Sayles (1988) suggests that the long-term behaviour of frozen ground can be likened to the frictional resistance of the ground itself, with the ice capable of sustaining stress. Frozen earth structures are essentially rigid, and so should be designed to resist "at rest" earth pressures.

The key reference on the practice of ground freezing is the important text of Harris (1995).

10.1.7 References

Gallavresi, F (1985)
"Soil improvement by means of ground freezing"
Symp on recent developments in ground improvement techniques, Bangkok, Dec 1982
AA Balkema, Rotterdam, pp 459–468

Gysi, H and Mader, P (1987)
"Ground freezing technique – its use in extreme hydrological conditions"
Proc 9th Eur Conf Soil Mech and Found Engg, Dublin, Vol 1, pp 165–173

Harding, H J B and Glossop, R (1939)
"Modern processes in the support of excavations"
The Engineer, 19 May

Harris, J S (1988)
"State of the art: tunnelling using artificially frozen ground"
Proc 5th Int Symp on Ground Freezing (R H Jones and J T Holden, eds)
AA Balkema, Rotterdam, 1988, pp 145–253

Harris, J S (1995)
Ground freezing in practice
Thomas Telford, London

Harris, J S and Pollard, C A (1985)
"Some aspects of groundwater control by the ground freezing and grouting methods"
Proc Reg Mtg Eng Grp of Geol Soc, Sheffield, September, 1985

Harris, J S, Holden, J T and Jones, R H (1992)
"A grounding in freezing"
Ground engineering, Vol 25, No 7, September, pp 42–45

Harvey, S J and Martin, C J H (1988)
"Construction of the Asfordby mine shafts through Bunter sandstone by use of ground freezing"
In: *Ground Freezing 88* (R H Jones and J T Holden, eds)
Proc 5th Int Symp on Ground Freezing, AA Balkema, Rotterdam, pp 339–348

Jessberger, H L (1985)
"The application of ground freezing to soil improvement in engineering practice"
Symp on recent developments in ground improvement techniques, Bangkok, Dec 1982
AA Balkema, Rotterdam, pp 459–468

Jessberger, H L (1992)
"Artificial freezing of the ground for construction purposes"
In: *Ground engineer's reference book* (F G Bell, ed)
Butterworths, London, pp 31/1–31/17

Kay, B D and Perfect, E (1988)
"State of the art: Heat and mass transfer in freezing soils"
Proc 5th Int Symp on Ground Freezing (R H Jones and J T Holden, eds)
AA Balkema, Rotterdam, 1988, pp 3–21

Klein, J (1988)
"State of the art: Engineering design of shafts"
Proc 5th Int Symp on Ground Freezing (R H Jones and J T Holden, eds)
AA Balkema, Rotterdam, 1988, pp 235–244

Rohde, J K G, Berggren, A-L and Aas, G (1987)
"Ground freezing as water sealing and ground improvement"
Proc 9th Eur Conf Soil Mech and Found Engg, Dublin, 1987
AA Balkema, Rotterdam, Vol 1, pp 235–240

Sayles, F H (1988)
"State of the art: mechanical properties of frozen soil"
Proc 5th Int Symp on Ground Freezing (R H Jones and J T Holden, eds)
AA Balkema, Rotterdam, Vol 1, pp 143–165

Skipp, B O and Hall, M J (1982)
Health and safety aspects of ground treatment materials
Report 95, CIRIA, London

10.2   ARTIFICIAL GROUND HEATING

10.2.1   Definition

Artificial ground heating is where heat is applied to the ground that dries and binds the soil particles together as a form of improvement. It is, therefore, a form of thermal stabilisation (Mitchell, 1982).

10.2.2   Principle

Heating fine-grained soils to more than 100°C causes drying and strength gains, if re-wetting is prevented. It is a temporary measure. Heating from 600°C to 1000°C gives a permanent increase in strength and decreases the soil's swelling capacity and compressibility.

Heating the ground to a temperature high enough to cause the necessary changes in soil properties is achieved mainly by the infiltration of compressed, heated air. Alternatively, it can be an effect of the incandescent products of combustion through the soil pores (Litvinov, 1960).

The process is dominated by two factors: the vaporisation of water on heating above 100°C, and the groundwater flow in and out of the zone being treated. Soils must be permeable to gas to allow removal of water vapour (Mitchell, 1982). Temperatures must be kept below soil fusion to prevent pore blockage.

10.2.3 Description

Heating has been applied almost exclusively in Russia and Eastern Europe and has been applied principally to loess (a fine-grained, wind-blown collapsible soil). Litvinov (1960) describes the method as involving burning various fuels in the soil being treated. The process of combustion takes place in sealed boreholes 100–200 mm in diameter, with control of temperature and chemical composition of the combustion products. Figure 10.6 shows the arrangement of a typical installation.

Fuels can be gaseous, liquid or solid, and are burnt at the mouth of the borehole or within the soil mass itself to give temperatures of between 750°C and 1000°C. An excess pressure of the hot gases of 0.25–0.50 atm is maintained above atmospheric pressure. This can increase gas permeability by three to five times. As there is no outlet, the incandescent gaseous products of combustion have to infiltrate through the soil pores. The temperature of the hot gases can be controlled by altering the amount of air blown into the borehole. This has to be between two-and-a-half and three times the minimum quantity required for complete fuel combustion. Treatment is carried out for eight to 10 days to form a consolidated zone about 2–3 m in diameter and typically 8–10 m deep. Between six and 30 boreholes can be treated simultaneously, with several cycles of treatment sometimes being used. Continuous control is needed. Treatment depths can extend to 20 m.

Alternatively, electric heaters can be used, with compressed air blown through the heater at the top of the borehole. This can produce temperatures of 500–1200°C. Electric heaters can also be lowered down the boreholes (Mitchell, 1982).

10.2.4 Applications

Beles and Stanculescu (1958), Litvinov (1960) and Mitchell (1982) report the use of heating applied to loess soils to stabilise landslides, underpin buildings, reduce lateral stresses on walls, and to form foundations. Table 10.4 gives guidance on the minimum temperatures required for specific tasks. Heating can be used to sinter the ground around the borehole to form vitrified piles.

Table 10.4 *Minimum heat treatment temperatures for different applications (from Mitchell, 1982)*

Purpose of heating	Minimum treatment temperature (°C)
Reduction of lateral pressure	300–500
Elimination of collapse properties (loess)	350–400
Control frost heave	500
Massive column construction below frost depth	600
Manufacture of building materials	900–1000

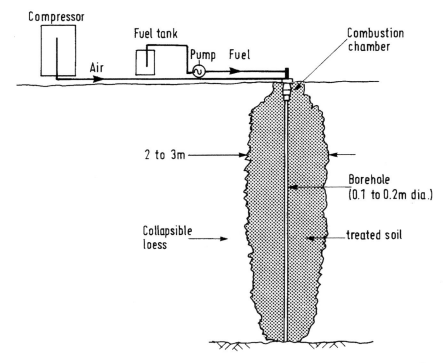

Figure 10.6 *Typical combustion installation for artificial ground heating (after Litvinov, 1960)*

10.2.5   Limitations

The obvious limitation for heating is the high energy cost of gaseous, solid or liquid fuels, using either combustion or electricity, to carry out the process. The high price of fuel may condemn the process to obscurity (Ingles, 1982).

10.2.6   Design

In view of the limitations on the process, it is felt not useful to give any guidance on design. Costings can be assessed using the proposals of Ingles and Metcalf (1972), summarised by Mitchell (1982).

10.2.7   The effect of heat generated as a by-product

In contrast to the deliberate application of large amounts of heat energy described above, the addition of lime and cement to soil and the chemical reactions they produce generates relatively low amounts of heat that can have a beneficial effect. In lime columns, for example, initial rapid rises in temperature of 50–200°C gradually reduced over two weeks to the ambient ground temperature. Such temperatures are not reached in laboratory experiments. The effect of higher temperature on stabilised soils is to give a more rapid gain of the strength and stiffness. In some circumstances, the heat of hydration can be a disadvantage. Jet grouting in organic clays in Singapore to provide support for tunnel excavation caused very high temperatures and generated ammonia (Bell, 1988).

10.2.8 References

Beles, A A and Stanculescu, I I (1958)
"Thermal treatment as a means of improving the stability of earth masses"
Géotechnique, Vol 8, pp 158–165

Bell, A L (1988)
"Report of discussion on Coomber, D B on Jet Grouting"
Proc Instn Civ Engrs, Pt 1, Vol 80, pp 1661–1664

Ingles, O G (1982)
"Thermal stabilization, discussion in Session 12 Soil improvement"
Proc 10th Int Conf Soil Mech and Found Engg, Stockholm, 1981
AA Balkema, Rotterdam, Vol 4, pp 968–969

Ingles, O G and Metcalf, J B (1972)
Soil stabilisation
Butterworths, Sydney

Litvinov, I M (1982)
"Stabilization of settling and weak shaley soils by thermal treatment"
In: *Soil and Foundation Engineering*
Special Report 60, Highway Research Board, Washington, DC, pp 94–112

Mitchell, J K (1982)
"Soil improvement – state of the art"
Proc 10th Int Conf Soil Mech and Found Engg, Stockholm, 1981
AA Balkema, Rotterdam, Vol 4, pp 509–565

11 Improvement by vegetation

11.1 DEFINITION

Vegetation as ground improvement is the biological reinforcement of ground by plant roots to retain earth masses and prevent soil loss. It is a combination of engineering and horticulture (Gray and Leiser, 1982).

11.2 PRINCIPLE

Vegetation can act as a means of ground improvement by the following mechanisms:

- reinforcement through the development of the root network
- removal of soil water by evapo-transpiration
- attenuation of the erosive forces of wind and rain
- trapping soil particles that have been eroded.

The mechanical reinforcement provided by roots, because of their individual and combined tensile strength and spread in the ground, can increase apparent cohesion considerably. Barker (1986) suggests that a root content of 1–2 per cent can increase cohesion by two or three times. The effect is generally shallow, as 80–90 per cent of all roots are less than 1 m deep. Deeply rooted trees, however, can act by a combination of buttressing and soil arching in much the same way as slope dowels (Section 7.4) anchoring the surface soils at depth (Gray, 1978).

The natural removal of soil moisture in the evapo-transpirative process tends to increase the soil strength, particularly in clayey soils, and may reduce the rate of infiltration by rainwater through the soil. High soil suctions (negative pore pressures) can be generated. Clay soils shrink in volume as moisture is removed.

As well as the binding action of the root network, the vegetation can also act to lessen the erosive power of wind or rain, eg in breaking the fall of raindrops or reducing the velocity of run-off water or winds at the ground surface. In this respect, the vegetation can create conditions where soil particles will settle out of their suspension in wind or run-off and become trapped – the principle by which coastal sand dunes are stabilised with marram grass.

11.3 DESCRIPTION AND APPLICATIONS

The use of vegetation in civil engineering is much wider than ground improvement, as it can have an engineering, landscape or other environmental function. For a comprehensive exposition of the subject see Coppin and Richards (1990), which provides guidance on the selection, design and management of vegetation in engineering projects.

The commonest use is the grassing of soil slopes to increase surface stability as well as appearance. The technique of hydroseeding (spreading seeds in a sprayed water slurry containing fertiliser, mulch and a binder) is a rapid method of seeding large areas and steep slopes. This, and other methods of seeding, can be combined with the use of biodegradable or permanent geotextiles (see Figure 11.1).

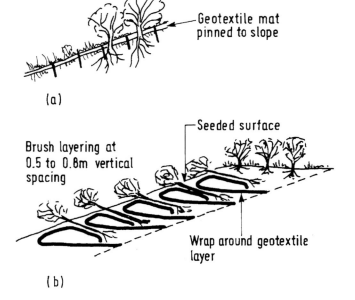

(a)

Brush layering at 0.5 to 0.8m vertical spacing

Seeded surface

Wrap around geotextile layer

(b)

Figure 11.1   *Slope improvement with vegetation: (a) planting and seeding with surface geotextile mat, (b) wrap-around geotextile and brush layering*

Trees and shrubs are used in various ways to help slope stability by resisting shallow rotational and translational slides within the upper 1–2 m of the slope surface. Again they can be combined with other techniques of slope reinforcement, eg gabion buttresses, geotextiles and crib walls.

Table 11.1 from Gray and Leiser (1982) summarises the approach to slope protection and erosion control and the roles vegetation can play.

11.4   LIMITATIONS

In all cases, vegetation takes a considerable time to become established. Table 11.2, based in part on a similar table from the Hong Kong Geotechnical Control Office lists benefits and disadvantages of vegetation in relation to slope stability. In Hong Kong, the torrential rains are not infrequently associated with typhoons, so uprooted trees create greater risk of slope failure by opening it up to the rain.

The success of vegetation depends on selecting appropriate cultivars for the conditions and can be affected by:

- its location, eg orientation of the slope face
- the soil type and subsoil character
- blight or drought
- salt water (and spray)
- air and water pollution
- grazing by animals
- pedestrians.

Table 11.1   *Approaches to slope protection and erosion control*

	Category	Examples	Appropriate uses	Stabilising mechanism or role of vegetation
Live construction	Conventional plantings	Grass seeding Transplants	Control of surficial rainfall and wind erosion To minimise frost effects	To bind and restrain soil particles To filter soil from runoff To intercept raindrops To maintain infiltration To change thermal character of ground surface
Mixed construction	Woody plants used as reinforcement and as barriers to soil movement	Live staking Contour-wattling Brush-layering Reed-trench-terracing Brush mats	Control of surficial rainfall erosion (rilling and gullying) Control of shallow (translational) mass movement	Same as above, but also to reinforce soil and resist downslope movements of earth masses by buttressing and soil arching action
	Woody plants grown in interstices of low porous structures or benches of tiered structures	Vegetated revetments (riprap, grids, gabion mats, blocks) Vegetated retaining walls (open cribs, gabions, stepped-back walls, and welded-wire walls)	Control of shallow mass movements and resistance to low-mod earth forces Improvement of appearance and performance of structures	To reinforce and indurate soil or fill behind structure into monolithic mass To deplete and remove moisture from soil or fill behind structure
	Toe walls at foot of slope used in conjunction with plantings on the face	Low breast walls (stone, masonry, etc) with vegetated slope above (grasses and shrubs)	Control of erosion on cut and fill slopes subject to undermining at the toe	To stop or prevent erosion on slope face above retaining wall
Inert construction	Conventional structures	Gravity walls Cantilever walls Pile walls Reinforced earth walls	Control of deep-seated mass-movement and restraint of high lateral earth forces Retention of toxic or aggressive fills and soil	Mainly decorative role

11.5   DESIGN

Reference should be made to Coppin and Richards (1990) for guidance on the design of engineered vegetation systems. While their purpose is almost invariably for long-term, rather than temporary, improvement, climatic, human and other environmental factors can rapidly cause deterioration or complete destruction. Thus, it would be wrong to rely on any contribution from vegetation to provide slope stability.

Table 11.2   *Effects of vegetation on slope stability*

Benefits
- interception of rainfall by foliage, reducing surface erosion
- reduction of evaporative losses of soil moisture at ground surface
- depletion of soil moisture and increase of soil suction
- mechanical reinforcement by roots
- restraint by buttressing and soil arching between tunnels of trees
- arresting the roll of loose boulders

Disadvantages
- maintaining infiltration capacity
- root wedging of near-surface rocks and boulders
- susceptibility to wind uprooting and exposing or loosening the ground
- surcharging the slope by large heavy trees (in some cases this could have a benefit)
- shrinkage and cracking of clay soils
- root interference with slope drainage systems
- susceptible to fire

11.6   REFERENCES

Barker, D H (1986)
"The enhancement of slope stability by vegetation"
Ground Engineering, April, pp 11–15

Coppin, N J and Richards, I G (eds) (1990)
Use of vegetation in civil engineering
Book 10, CIRIA, London/Butterworth, London, 292 pp

Gray, D H (1978)
"Role of woody vegetation in reinforcing soils and stabilizing slopes"
Proc Symp Soil reinforcing and stabilizing technique in engineering practice
NSW Inst Tech, Sydney, Australia, pp 253–306

Gray, D H and Leiser, A J (1982)
Biotechnical slope protection and erosion control
Van Nostrand Reinhold, New York

12 Options for choice

Vibro-replacement (stone, and now concrete, columns) has been used for almost 90 per cent of the ground improvement schemes carried out for building developments in the UK, with dynamic compaction the next most frequent. Pre-compression and deep drains are used occasionally, mostly for embankments to roads, although pre-compression is sometimes applied to housing and industrial sites. The uses of geotextiles, reinforced earth are well established; jet grouting and soil mixing are also developing rapidly. On civil engineering projects, the well-established techniques of dewatering, grouting, and freezing are called upon as temporary works to enable the permanent works to be constructed more easily. To an increasing extent, the newer techniques are also becoming part of the permanent works, eg geotextiles, embankment piles, although the instances are not numerous.

Many of the ground treatment methods have been used for more than 40 years. However, as Greenwood (1984) said of the "art of compaction", they remain largely empirical. Although there have been some advances recently in design theories, the difficulties in accurately determining the properties and fabric of soils still inhibit progress. The same applies to deep drains, for which Hansbo *et al* (1982), Akagi (1982) and Holtz *et al* (1991) note that there are many uncertainties, even though predictive theory has long been available (ie Barron, 1948). It is prudent before deciding on a method of ground treatment to examine the case histories of its use for similar ground conditions. *It is also prudent not to apply processes near the limits of their usual range of application.*

The designer with a problem site looks for assistance, ideas, references, warnings, guidance – in other words, previous experience – to help towards a solution. Published comparisons of techniques used on the same site are helpful, but only to a limited extent. This is because almost all have been about deep compaction techniques on sites of loose or silty sands, often where seismicity had to be considered. However, from these references some fairly clear guidelines emerge (see Appendix B).

Some examples of the deliberate combination of several methods targeted to improve different features of the ground at a site are mentioned in Section 12.1. The choice of process (Section 12.2) can often be examined in relation to cost-effectiveness compared with piling, and from the point of view of the limitations of each process. The available time for construction might be a determining factor in choosing a method of treatment or whether to pursue a treatment option at all.

12.1 COMBINATIONS OF TECHNIQUES

In certain circumstances, a combination of techniques may be appropriate to satisfy different aspects of, or objectives for, ground improvement. While combinations such as pre-loading with vertical drains, or phases of grouting progressively by different methods, have long been normal practice, the underlying purpose is essentially the same as that of the component methods.

Increasingly, it is possible to choose the technique best suited to solve one engineering problem and other techniques to overcome separate or related problems. An example of this is using reinforced soil (to reduce earth fill volumes or the land-take for side slopes) and strengthening the foundation soils for stability or to minimise or equalise differential settlements by jet grouting, lime columns, vibratory compaction or vertical drains.

The permutations that have become possible open up many opportunities for faster construction times, savings of space and volume of imported fills, and – although this is a matter of engineering judgement – compatibility of soil and structure behaviour. Not least of the advantages is the ability to choose methods suitable for the different soil types or depths at which they are present.

One combination that has proved particularly successful over the past 30 years is the use of pre-loading (Section 5.1) with vertical drains (Section 6.2). To determine the consolidation characteristics of the ground, necessary for the design of the vertical-drain scheme, it is desirable, certainly for major projects, to monitor the performance of a trial embankment. It is also usual to take advantage of the trial to compare the performance of different types of deep drains in accelerating settlement (Holtz et al 1991). Typical examples of these trials in the United Kingdom are reported by Nicholson and Jardine (1982), in Hong Kong by Foott et al (1987) and in Spain by Abella et al (1988).

Extended combinations of techniques are *pre-loading and vertical drains, with geotextiles* (Ting et al 1990), and also using lightweight fill (Seim et al, 1981). In addition, *vacuum pre-loading and deep drains* have been combined, as Kjellman (1952) suggested that they should be. Examples are described by Ye et al (1984), Chen and Bao (1984), and Sehested and Yee (1990).

Geotextiles are employed commonly with *embankment piles* or *vibrated concrete columns* to create load-transfer platforms between the isolated pile caps. Holtz and Massarsch (1976) and Collingwood and Fenwick (1985) describe examples. *Geotextiles* can also be used to wrap stone columns (ie *vibro-replacement*) where the surrounding ground provides poor restraint to the column.

Inundation and blasting were used in conjunction in Russia (Litvinov, 1973) and in Bulgaria (Donchev, 1980).

Reinforced soil structures have been combined with *vibro-replacement* to try to ensure compatibility of movement between the foundations and the reinforced soil wall itself. Examples are described by Di Maggio and Goughnour (1979), Barksdale and Goughnour (1984), Munfakh (1984), and Munfakh et al (1984). Alternatively, a rigid foundation can be provided; for example a *reinforced soil* wall was supported on *micro-piles* (Miki and Kodama, 1985).

The unusual combination of *freezing, jet grouting and vibro-compaction* for the construction of a tunnel below the River Limmat in Switzerland is described by Gysi and Mader (1987). This example is mentioned in Section 10.1.4.

Probably the most frequently combined methods (more so than preloading and deep drains) is *vibro-replacement and dynamic compaction*. Solymar and Reed (1986), Keller et al (1987), and Hussin and Ali (1987) describe typical examples. Slocombe (1989) describes a UK example of a derelict site where, in addition to these two methods, piling and grouting were also used. For the lighter structural loads and because of the extent of alluvial clays, vibro-replacement and dynamic compaction complemented each other. The grouting was necessary to stabilise old mine workings, while the piles, provided to support the heaviest building loads, were also used to form retaining walls.

Useful recent references about the use of techniques in combination are Mitchell and Welsh (1989) and Choa (1989).

12.2　DISCUSSION ON CHOICE OF METHODS

The choice of method or methods is the essential step in the design process (see Sections 1.3 and 3.3). The designer's choice has to take account of the possible hazards of the site in relation to the engineering options. It is strongly recommended that these are discussed fully with the promoter to reach agreement on the use of ground improvement.

Among the factors to be considered are risks associated with the method of treatment:

- noise and vibration effects

- temporary lowering of the groundwater

- horizontal and vertical deformations, eg displacement methods can cause substantial ground heave – pile installations often being the most serious (van Weele, 1984), or settlements can extend outside the site

- uncertainties about, or deterioration of, the performance of the materials installed in the ground, eg the discharge capacity of vertical drains being reduced by clogging, folding or siltation (Holtz *et al*, 1991)

- changes in lateral effective stresses as, eg, when the ground expands on freezing

- dealing with the requirements of environmental legislation and regulation, eg discharge consents and waste management.

A critical factor in the choice of method is time: when? and how long? Should the ground treatment be carried out before the construction of the main works, during construction, or is it part of a scheme of remedial or reconstruction work? Hartikainen (1984) discussed this point in relation to specific types of improvement process. Is the improvement to be achieved by the treatment an intrinsic part of the permanent works or a construction expedient (and hence thought of as temporary works – the contractor's responsibility) or, possibly, having both purposes?

This last point about temporary works is particularly important in relation to underground construction. The designers of the project may foresee the need for ground treatment, but the choice and amount of treatment is often left to the main contractor. The design of the treatment is usually a matter for a specialist sub-contractor. Should the treatment prove inadequate or, worse, inappropriate – and many shafts and tunnel projects of the past bear witness to this – there can be serious consequences of increased costs and delay. Often, these situations have arisen despite willingness, effort and good-quality work. There are obvious needs for better knowledge about the ground that has to be treated and clear thinking about the responsibilities for design in relation to the promoter. But an important element is allowing time for the treatment methods to evolve as knowledge about the ground builds up during the initial stages. This, perhaps, is where there is a difference of approach between ground treatment for temporary works and for permanent works.

It should again be emphasised that many ground improvement techniques are sensitive to small variations in the properties of the ground. It is essential, therefore, not to apply a process beyond its limits. If ground conditions vary so that this happens, the consequences could involve:

- intensifying the treatment

- increasing the expenditure of energy

- increasing the time for pre-consolidation or stabilisation

- reducing the acceptance criteria, signifying a modified structure and/or foundation

- bringing in an entirely different technique.

It is far better not to go too close to the suitability limit for the process (Hartikainen, 1984). The designer should ensure that the specialist contractor also understands:

- the design and the structure proposed
- the nature and properties of the ground.

Hansbo (1984) pointed out that it really depends where you are in the world for a ground improvement option to be cost effective against piling. In order to illustrate the point Table 12.1 shows in very broad terms the relative costs in the UK for a vibro-replacement solution compared with piling in relation to the probable relative order of settlement. Ground treatment, therefore, opens up a choice; of options for the designers of the whole scheme, and to the promoter, whose money pays for it, of savings.

Table 12.1   *Relative order of costs and settlements for foundation options*

	Do nothing	Ground improvement eg vibro-replacement	Piled foundation
Relative settlement	1	½	1/10–1/20
Relative cost (inc substructure)	0	1–2	2–3

With other types and purposes of ground improvement, savings are made possible by preplanned applications. Allowing sufficient time for pre-loading to induce settlements, for example, could eliminate the need for embankment piles.

On the other hand, and usually even more important in the economics of development, is earlier occupation and use of the site; in this respect ground improvement has an important role.

12.3   REFERENCES

Abella, S, Rivas, F and Velasco, M (1988)
"Shallow foundations on improved collapsible soils"
In: *Shallow foundations*
Geotechnical Special Publication 15, Am Soc Civ Engrs, pp 17–26

Akagi, T (1982)
"Effects of mandrel-driven sand drains on soft clay"
Proc 10th Int Conf on Soil Mech and Found Engg, Stockholm, Sweden, 1981
AA Balkema, Rotterdam, Vol 3, pp 581–584

Barksdale, R D and Goughnour, R D (1984)
"Settlement performance of stone columns in the US"
Proc Int Conf In-situ soil and rock reinforcement, Paris
Presses Points et Chaussées, pp 105–110

Barron, R A (1948)
"Consolidation of fine-grained soils by drain wells"
Trans Am Soc Civ Engrs, Vol 113, Paper 2346, 718–754

Chen, H and Bao, X-C (1983)
"Analysis of soil consolidation stress under the action of negative pressure"
Proc 8th Eur Conf Soil Mech and Found Engg, Helsinki, 1983
AA Balkema, Rotterdam, Vol 2, pp 591–596

Choa, V (1989)
"Drains and vacuum preloading test"
Proc 12th Int Conf Soil Mech and Found Engg, Rio de Janeiro, Vol 2, pp 1347–50

Collingwood, R W and Fenwick, T H (1985)
"Selby diversion of the East Coast Main Line: Construction"
Proc Instn Civ Engrs, February, Vol 77, Part 1, pp 49–84

Di Maggio, J A and Goughnour, R D (1979)
"Demonstration program on stone columns"
In: *Soil reinforcement*, Proc Int Conf, Paris
Edition Anciens, Ecole National des Ponts et Chaussées, 249–254

Donchev, P (1980)
"Compaction of loess by saturation and explosion"
Proc Colloque Int sur le compactage, Paris, Vol 1, pp 313–317

Foott, R, Koutsoftas, D C and Handfelt, L D (1987)
"Test fill at Chek Lap Kok"
J Geotech Engg, Am Soc Civ Engrs, Vol 113, No 2, pp 106–126

Greenwood, D A (1984)
"Summary of discussion on deep compaction"
Proc 8th Eur Conf Soil Mech and Found Engg, Helsinki, 1983
AA Balkema, Rotterdam, Vol 3, pp 1131–1132

Gysi, H and Mader, P (1987)
"Ground freezing technique – its use in extreme hydrological conditions"
Proc 9th Eur Conf Soil Mech and Found Engg, Dublin
Vol 1, pp 165–173

Hansbo, S (1984)
"Techno-economic trend of subsoil improvement methods in foundation engineering"
Special lectures, *Proc 8th Eur Conf Soil Mech and Found Engg, Helsinki, 1983*
AA Balkema, Rotterdam, Vol 3, pp 1333–1342

Hansbo, S, Jamiolkowski, M and Kok, L (1982)
"Consolidation by vertical drains"
In: *Vertical drains*, Thomas Telford, London, pp 45–66

Hartikainen, J (1984)
"On the geotechnical design of foundations in improved subsoil"
Proc 8th Eur Conf Soil Mech and Found Engg, Helsinki, 1983
AA Balkema, Rotterdam, Vol 3, pp 1319–1332

Holtz, R D, Jamiolkowski, M B, Lancellotta, R and Pedroni, R (1991)
Prefabricated vertical drains: design and performance
Book 11, CIRIA, London/Butterworth-Heinemann, Oxford

Holtz, R D and Massarsch, K R (1976)
"Improvement of the stability of an embankment by piling and reinforced earth"
Proc 6th Eur Conf on Soil Mech and Found Engg, Vienna, Vol 1, pp 473–478

Hussin, J D and Ali, S (1987)
"Soil improvement at the Trident Submarine facility"
In: *Soil improvement – a ten year update* (J P Welsh, ed)
Geotechnical Special Publication 12, Am Soc Civ Engrs, pp 215–231

Keller, T O, Castro, G and Rogers, J H (1987)
"Steel Creek dam foundation densification"
In: *Soil improvement – a ten year update* (J P Welsh, ed)
Geotechnical Special Publication 12, Am Soc Civ Engrs, pp 136–166

Kjellman, W (1952)
"Consolidation of clay soil by means of atmospheric pressure"
Proc Conf on soil stabilisation, MIT, pp 258–263

Litvinov, I M (1973)
"Deep compaction of soils with the aim of considerably increasing their carrying
capacity – discussion"
Proc 8th Int Conf on Soil Mech and Found Engg, Moscow, Vol 4, pp 365–367

Miki, G and Kodama, H (1985)
"Practical uses of the root pile method in Japan"
Proc Symp Recent developments in ground improvement techniques, Bangkok, 1982
AA Balkema, Rotterdam, pp 433–438

Mitchell, J K and Welsh, J P (1989)
"Soil improvement by combining methods"
Proc 12th Int Conf Soil Mech and Found Engg, Rio de Janeiro
AA Balkema, Rotterdam, Vol 2, pp 1293–1296

Munfakh, G A (1984)
"Soil reinforcement by stone columns – varied case application"
Proc Int Conf on in-situ soil and rock reinforcement, Paris, pp 157–162

Munfakh, G A, Sarkar, S K and Castelli, R J (1984)
"Performance of a test embankment founded on stone columns"
In: *Piling and ground treatment*, Thomas Telford, London, pp 259–265

Nicholson, D P and Jardine, R J (1982)
"Performance of vertical drains at Queensborough Bypass"
In: *Vertical drains*, Thomas Telford, London, pp 67–90

Sehested, K G and Yee, T S (1990)
"Soil improvement using vertical band drains and vacuum preloading of Section 6/7"
Proc Int Symp on trial embankments on Malaysia Mance Clays, Kuala Lumpur
(R R Hudson, C T Toh and S F Chan, eds)
Malaysian Highway Authority, Vol 2, pp 89–101

Seim, C, Walsh, T J and Hannon, J B (1981)
"Wicks, fabrics and sawdust overcome thick mud"
Civil Engineering, Am Soc Civ Engrs, July, pp 53–56

Slocombe, B C (1989)
"Thornton Road, Lister Hills, Bradford"
In: *Piling and deep foundations* (J B Burland and J M Mitchell, eds), Proc Int Conf, London
AA Balkema, Rotterdam, Vol 1, pp 131–142

Solymar, Z V and Reed, D J (1986)
"A comparison of foundation compaction techniques"
Canad Geotech J, Vol 23, No 3, August, pp 271–280

Ting, W H, Chan, S F and Kaisson, K (1990)
"Embankments with geogrids and vertical drains
Proc Int Symp on trial embankments in Malaysian Marine Clays, Kuala Lumpur
(R R Hudson, C T Toh, and S F Chan, eds)
Malaysian Highway Authority, Vol 2, pp 35–47

Van Weele, A F (1984)
"Soil improvement methods in a dense urban environment"
Proc 8th Eur Conf Soil Mech and Found Engg, Helsinki, 1983
AA Balkema, Rotterdam, Vol 1, pp 1345–1354

Ye, B-R, Lu, S-Y, and Tang, Y-S (1984)
"Packed sand drain – atmospheric preloading for strengthening soft foundations"
Proc 8th Eur Conf Soil Mech and Found Engg, Helsinki, 1983
AA Balkema, Rotterdam, Vol 2, pp 712–720

Appendices

A1 Settlement of fills

A1.1 UNDER SELF-WEIGHT (CREEP SETTLEMENT)

This continuing settlement takes place even though there is no change of stress or moisture content. Its amount and rate can be estimated using the parameter α, which is defined by Charles (1984) as percentage vertical compression of the fill occurring in a log cycle of time, eg between one year and 10 years after placement. Typical values given by Charles are shown in Table A1.1.

This estimation method applies only when the character of the ground and loading conditions remain unaltered.

Table A1.1 *Settlement of fills under self-weight*

Type of fill	α (per cent)
Well-graded rock fill	0.2
Uncompacted opencast mining fill	0.5–1.0
Domestic refuse with high organic content	2–10 (depending on rate of decay or decomposition)
Old domestic refuse	1.0

A1.2 UNDER INCREASED LOAD

When additional load is placed on fills, Charles (1984) suggests that one-dimensional (vertical) compression can be estimated using a value of constrained modulus.

Constrained Modulus, $D = \Delta\sigma_v / \Delta\varepsilon_v$ (Lambe and Whitman, 1979)

where $\Delta\sigma_v$ is the increment of vertical stress and $\Delta\varepsilon_v$ is the corresponding change in vertical strain.

This approach applies to values of $\Delta\sigma_v$ up to about 100 kN/m^2 where the initial vertical stress is about 30 kN/m^2. Typical values of constrained modulus, D, for several types of fill are given in Table A1.2. The time rate at which this settlement happens depends on the nature of the fill. Much of the settlement with loose unsaturated fills, and even of lagoon-deposited pfa, takes place as the load is placed on the fill.

A1.3 BECAUSE OF INUNDATION

Loose unsaturated fills can settle suddenly (collapse settlement) when inundated with water. This can be caused by rainwater entering the fill, possibly through backfilled excavations for drains or soakaways, or by a rise in groundwater levels. The possibility of stone columns acting as water entry channels should be recognised as part of the design of the ground improvement. Similarly, a rise in water table may cause collapse settlement following cessation of dewatering or deep pumping.

One situation to be considered might be a rise in the groundwater levels below cities because of the reduced pumping for water supply (see Simpson *et al*, 1989, for the engineering implications of the rise in water levels below London). Another situation is leakage from water mains.

Table A1.2   *Compressibility of untreated fills*

Fill type	Compressibility	Typical value of constrained modulus (kN/m^2)
Dense well-graded sand and gravel	very low	40 000
Dense well-graded sandstone rockfill	low	15 000
Loose well-graded sand and gravel	medium	4000
Old urban fill	medium	4000
Uncompacted stiff clay fill above water table	medium	4000
Loose well-graded sandstone rockfill	high	2000
Poorly compacted colliery spoil	high	2000
Old domestic refuse	high	1000–2000
Recent domestic refuse	very high	<1000

There are several examples related particularly to colliery waste backfill of open-cast mines (Charles *et al*, 1984; Smyth-Osbourne and Mizon, 1984; Charles and Burford, 1985; Schulz *et al*, 1986). Charles (1984) reports collapse compressions of 2–6 per cent.

Chalk fills are also susceptible to collapse settlement (Clayton, 1980 and Stroud and Mitchell, 1990). Settlement in newly placed or newly loaded chalk fill usually occurs with heavy autumnal rain, following a dry summer, and the effects can be aggravated by stone columns acting as drains.

A1.4   BY CONSOLIDATION

When material is placed under water, as in many tailings lagoons, a soft cohesive-like fill is formed. Hughes and Windle (1976) and Ball (1979) suggest that the deposited material ground can be considered to behave as a soft clay, ie that the settlement amount and timing can be estimated by conventional consolidation theory. A separate consideration for very loose saturated fills is their potential for liquefaction. Some ground improvement methods are appropriate for reducing this potential (Bell *et al*, 1987), although it should be appreciated that they may actually induce liquefaction so as to create a denser fill. The settlement of loose or soft natural soils in relation to ground improvement can also be estimated by soil mechanics principles. Eggestad (1984) summarises ground improvement in cohesive soils, and Brandl (1984) in cohesionless soils.

A1.5   REFERENCES

Ball, M J (1979)
"The investigation of a quarry waste salt lagoon on the M42 Motorway"
Proc Symp on Engineering behaviour of industrial and urban fill
Birmingham, Midland Geotechnical Society, pp C11–24

Bell, A L, Slocombe, B C, Nesbitt, A M and Finey, J T (1987)
"Land redevelopment involving ground treatment by dynamic compaction"
In: *Building on marginal and derelict land*
Thomas Telford, London, pp 791–797

Brandl, H (1984)
"Improvement of cohesionless soils"
Proc 8th Eur Conf Soil Mech and Found Engg, Helsinki, 1983
AA Balkema, Rotterdam, Vol 3, pp 1009–1026

Charles, J A (1984)
"Settlement of fill"
In: *Ground movements and their effects on structures* (P B Attewell and R K Taylor, eds)
Surrey University Press, Guildford, pp 26–45

Charles, J A, Hughes, D B and Burford, D (1984)
"The effect of a rise of water table in the settlement of backfill at Horsley opencast coal mining site, 1973 to 1983"
In: *Ground movement and structures* (J D Geddes, ed)
Pentech Press, London, 1985, pp 423-42

Charles, J A and Burford, D (1985)
"The effect of a rise of water table on the settlement of opencast mining backfill"
Information Paper IP15/85, Building Research Establishment, Garston

Clayton, C R I (1980)
"The collapse of a compacted chalk fill"
Proc Colloque Int sur le Compactage, Paris, Vol 1, pp 119–124

Lambe, T W and Whitman, R V (1979)
Soil Mechanics
John Wiley and Sons, New York

Eggestad, A (1984)
"Improvement of cohesive soils"
Proc 8th Eur Conf Soil Mech and Found Engg, Helsinki, 1983
AA Balkema, Rotterdam, Vol 1, pp 991–1007

Hughes, J M D and Windle, D (1976)
"Some geotechnical properties of mineral waste tailings lagoons"
Ground Engineering, Vol 9, No 1, pp 23–28

Schultz, T M, Sonneberg, R G and Thomson, S (1986)
"A study of urban restoration of surface mined land in Western Canada"
In: *Building on marginal and derelict land*
Thomas Telford, London, pp 125–136

Simpson, B, Blower, T, Craig, R N and Wilkinson, W B (1989)
The engineering implications of rising groundwater levels in the deep aquifer below London
Special Publication 69, CIRIA, London

Smyth-Osbourne, J R and Mizon, D J (1984)
"Settlement of a factory on opencast backfill"
In: *Ground movements and structures* (J D Geddes, ed)
Pentech Press, London, pp 463–479

Stroud, M A and Mitchell, J M (1990)
"Collapse settlement of old chalk fill at Brighton"
In: *Chalk*, Proc Int Chalk Symp, Brighton, September 1989
Thomas Telford, London, pp 343–350

A2 Comparative case histories

A2.1 COMPARISONS OF TECHNIQUES

The results of case histories and comparative studies of different techniques reported here are taken from published material using the conclusions of their authors. Inevitably, and particularly because of the nature of these forms of ground engineering, there will be variations between the accuracy or quality of the case histories: few would meet the rigorous standards of engineering scientists. Neither CIRIA nor the authors have checked the source data to authenticate the results or to explain anomalies. Nevertheless, the compilation of previous experience in a form that allows comparison enables engineers to judge what can realistically be achieved. Always, of course, the judgement has to consider the differences between the reported work and the project being proposed, differences in ground conditions being the obvious, but not the only, ones.

There are relatively few case histories where different techniques have been used on the same site. The first, fundamental review of ground treatment processes was by Mitchell (1970). As part of his work Mitchell compared vibro-compaction, blasting and compaction piles.

The comparison is summarised in Table A2.1, which includes reasons for choice. Choices will always be influenced strongly by ground conditions, the expectation of achieving the required degree of improvement, and cost. Relative costs of different techniques, however, are not always consistent. Four case histories in Table A2.1 show vibro-compaction to be more economical than compaction piles. Three further case histories in the table indicate exactly the reverse. Mitchell (1970) summarises the findings as below.

1   Blasting and vibro-compaction should be limited to cohesionless soils with less than 20 per cent fines (silt) content (but see below for further discussion of this).

2   Compaction piles can be used with a somewhat higher silt content.

3   For blasting and compaction piles the soil has to be saturated. This is not necessary for vibro-compaction.

4   Vibro-compaction and blasting can be applied to depths of 30–40 m. Compaction piles may have an effective limit of about 15–20 m.

5   Vibro-compaction gave the most uniform improvement.

6   For suitable circumstances, blasting may be the fastest and cheapest method.

Basore and Boitano (1969) compared compaction piles and vibro-compaction in hydraulic fills. They found that vibro-compaction gave much higher and more consistent relative densities than compaction piles (Figure A2.1). It is interesting to note that although vibro-compaction was apparently cheaper, compaction piles were finally used for the project.

For saturated hydraulic fills, Schroeder and Byington (1972) combined published results with their own as shown in Figure A2.2. Vibro-compaction is shown as achieving higher relative densities than either the vibrating probe (Terraprobe) or compaction piles. Concern has been expressed, however, about the comparability of results obtained by various authors. Schroeder and Byington recommended using SPTs and surface level monitoring to control and check the effectiveness of the different treatment methods.

Table A2.1 *Some examples where alternative techniques of ground treatment were considered (after Mitchell, 1970)*

Project	Soil	Chosen method	Alternative considered	Reasons for choice
Paper mill foundations	Fine sand	Vibro-compaction	Compaction piles	Difficulty expected in driving piles and securing uniform compaction. Cost
Power plant foundation	Glacial sand and gravel	Vibro-compaction	Compaction piles	Cost and uniformity of compaction
Improvement below abutments	Not quoted	Vibro-compaction	Compaction piles	Compaction piles tried and found to be too slow
Fill a hole at a dam site	Uniform fine sand	Blasting	Vibro-compaction	Faster and reduced cost
Compressor station foundations	Loose fine and medium sand	Compaction piles	Vibro-compaction	Reduced cost
Dry dock foundations	Well-graded fill	Vibro-compaction	Blasting compaction piles	Risk of damage. Reduced cost
Support to reservoir base	Loose sand	Compaction piles	Vibro-compaction	Reduced cost
Building foundation	Loose sand with clay inclusions	Compaction piles	Vibro-compaction	Time schedule and fund availability

Note: in general the use of these methods was less expensive than having deep piled foundations

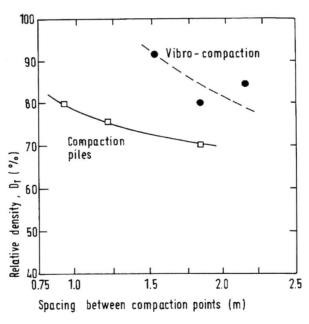

Figure A2.1 *Average relative densities and compaction point spacings of vibro-compaction and compaction piles (after Basore and Boitano, 1969)*

In comparing vibro-compaction with vibratory probing (Terraprobe) in densifying hydraulic fills, Brown and Glenn (1976) found that the penetration rate of the Terraprobe was four times that of the vibrator for vibro-compaction. However, four to five times as many treatment points were required for the Terraprobe to achieve the same relative density as vibro-compaction.

Maximum achieved relative density from vibro-compaction was always far higher than from the Terraprobe, regardless of probe spacing. Figure A2.3 shows the relationship of surface subsidence to the area per treatment point In this case, the higher degree of compaction obtained by the vibro-compaction is well illustrated.

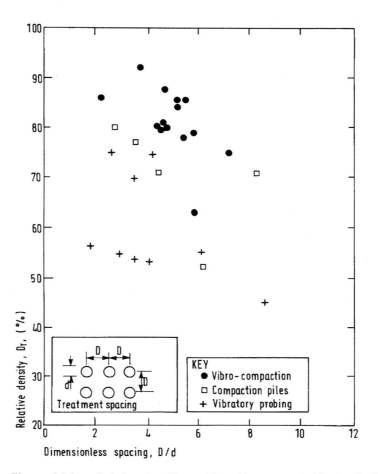

Figure A2.2 *Relative densities achieved in compacted hydraulic fills (after Schroeder and Byington, 1972)*

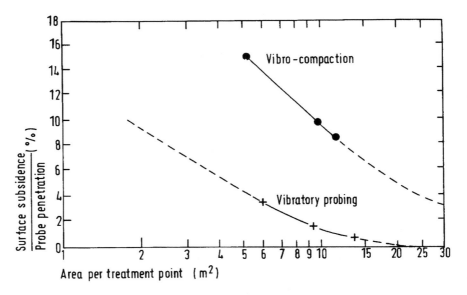

Figure A2.3 *Surface subsidence of hydraulic fill compacted by vibro-compaction and vibratory probing (after Brown and Glenn, 1976)*

Saito (1977) compared the vibro-rod, which he had developed, with vibro-compaction and with what is essentially a compaction pile, the vibro-compozer (Section 4.4.3), also in hydraulic fill. Figure A2.4 shows the result of the comparison. Saito suggests that vibro-compaction gave poor results because of the more than 15 per cent fines (silt) content. He illustrated the point with Figure 4.4. Although Mitchell (1982) referred to that plot in suggesting a limit of 20 per cent fines as a rule of thumb for the applicability of vibro-compaction, such a value is on the high side if there is to be substantial improvement. On the other hand, for the different type of process of the compozer system, the practicable range might extend to rather more than 20 per cent fines.

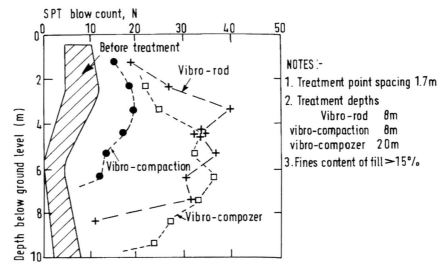

Figure A2.4   *SPT blow counts in hydraulic fill before and after three types of vibratory compaction (after Saito, 1977)*

A comparison between dynamic compaction, inundation and preloading, all applied to a cohesive fill with large air voids in a deep quarry, was made by Charles *et al* (1979). Table A2.2 lists the average settlements caused by each treatment and the relative costs. Houses were built on each of the treated areas, and Table A2.3 summarises the settlements during construction and at 10 years later (Charles *et al*, 1986).

Table A2.2   *Average settlements produced by ground treatment and treatment costs (Charles* et al, *1979)*

	Dynamic compaction	**Inundation**	**9 m-high surcharge**
At surface	240 mm	100 mm	410 mm
At 4 m depth	90 mm	40 mm	230 mm
At 10 m depth	<10 mm	< 10 mm	40 mm
Cost*	£15 000**	£2500	£12 000[+]

Notes

*   Cost of treatment of 50 m × 50 m square area at late 1974 prices.

** Dynamic compaction could have been considerably cheaper if a larger area had been treated.

[+]   Average haul distance was 100 m for placement of surcharge.

Table A2.3   *Settlement of experimental houses at Corby (Charles* et al, *1986)*

		Dynamic compaction	Inundation	Pre-loading (9 m surcharge)	Without treatment
Settlements (mm) during construction	Mean	7.0	6.1	1.4	2.7
	Maximum	9.2	14.3	3.0	6.8
	Minimum	3.2	2.8	-0.4	1.4
	Maximum differential	5.9	6.8	2.3	2.8
Total settlements (mm) during and subsequent to construction until 31 May 1984	Mean	38.8	41.9	7.8	19.3
	Maximum	49.2	126.3	17.6	33.0
	Minimum	18.3	16.1	3.0	8.9
	Maximum differential	20.3	79.9	6.0	11.1

Choa *et al* (1979) examined the effectiveness of dynamic compaction and the installation of deep drains in improving more than 40 m of soft marine clays. Dynamic compaction was not effective, but the deep drains were. Faraco (1981) reported trials of vibro-compaction (by two contractors) with the Terraprobe, compaction piles and blasting. The results are summarised on Figure A2.5 in which they are compared by SPT blow counts before and after treatment. Faraco noted that blasting gave poor results, but did not quote them. The contract used vibro-compaction and the Terraprobe.

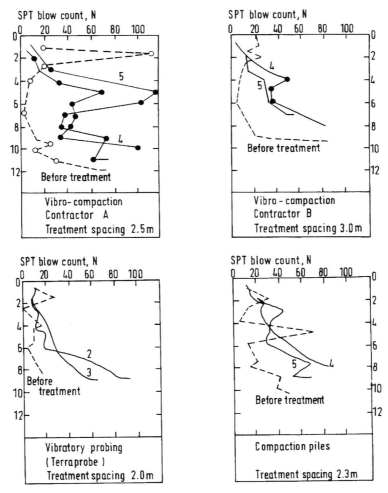

Figure A2.5   *SPT comparisons of vibratory compaction before and after different amounts of treatment*

Wallays (1985) used cone penetration testing to compare the Y-Probe (Section 4.5.3) with vibro-compaction, again applied to hydraulic fills. The results are shown on Figure A2.6. Wallays advised that if the fines content (particle sizes < 0.075 mm) is more than 12 per cent these processes are not suitable, except for the casing driver (a form of compaction pile – see Section 4.6). Murayama and Ichimoto (1985) report a comparison based on Swedish ram sounding between the vibro-compozer and vibro-compaction, in this case in "sandy ground" (Figure A2.7).

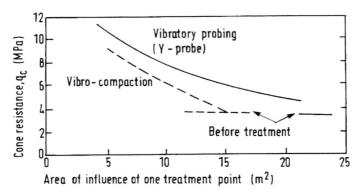

Figure A2.6   *Comparison of CPT cone resistances in hydraulic fill before and after vibratory probing and vibro-compaction (after Wallays, 1985)*

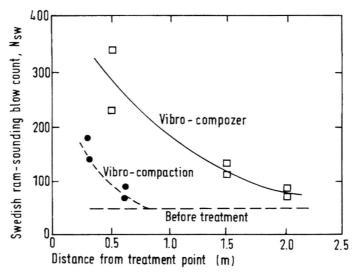

Figure A2.7   *Comparison of Swedish ram soundings in sandy ground before and after vibro-compaction in vibro-compozer compaction (after Murayama and Ichimoto, 1985)*

With all comparisons of techniques, the more so when a new method is shown in favourable light in contrast to an established method, it is necessary to examine the source references to check the basis of comparison and the details of the equipment and methods. Certainly, it would be unwise to draw general conclusions about the relative effectiveness of different techniques.

In a comprehensive comparison of techniques, Solymar and Reed (1986) describe three sites where vibro-compaction, dynamic compaction, compaction piles and blasting were used to reduce density variations in the ground. One objective was to increase the ground's lateral support for subsequently installed slender piles to prevent their buckling. Two of the three sites concerned both fine to medium sands and sand-gravel mixtures; the other had a range of soil types varying from gravelly fine sands to sandy or silty clays. The treatment layouts were developed by trials prior to contract. Each process was monitored using SPT and CPT soundings. Some of the conclusions that Solymar and Reed drew from their experience are as follows.

1   As well as detailed and suitable site investigation to choose the method, trials should be carried out to determine working procedures and to assess the suitability of equipment.

2   Strength increases, indicated by greater cone resistance, may continue over a period of days to months after compaction. This applies particularly to sands that have been subject to blasting. The timing of testing should allow for this phenomenon, which is discussed by Mitchell and Solymar (1984) and Mitchell (1986).

3   Dynamic compaction is not only suitable for sands, but it can also improve those that contain layers with some cohesive properties. The maximum practical depth that sand can be compacted is between 10 m and 20 m, depending on the relative density required.

4   Deep blasting is suited in partly to fully saturated clean loose sands, and saturated silts. The maximum depth of treatment was 45 m on one site. Relative densities of 65–70 per cent are considered to be the maximum consistently achievable.

5   Vibro-compaction is best suited to loose, clean, saturated fine sands. Depths of treatment extend to about 30 m. Density variations in the horizontal direction need to be considered for individual foundations. Relative densities of 70–80 per cent are attainable below 25 m, and higher values at shallower depths.

6   Compaction piling can be carried out in layered deposits using various types of equipment. A good source of backfill material is essential. Depths of more than 20 m can be reached, with relative density values of 75–80 per cent being possible.

7   The specified depth of improvement was most consistently achieved with vibro-compaction and deep blasting, with almost as good performance from compaction piling. Dynamic compaction showed the greatest variation with depth.

Keller *et al* (1987) were concerned about 12 m of layered loose clayey sands, and their liquefaction potential, beneath a dam. Test sections of dynamic compaction and vibro-replacement were compared by using three levels of compactive effort. Figure A2.8 shows the results from the two processes for the maximum effort imparted to the ground. Table A2.4 lists how some objectives were met and some were not. In the event, dynamic compaction was used for general densification while vibro-replacement was used in specific areas. Keller *et al* (1987) concluded that these treatments would not improve ground with 10 per cent or more clayey fines.

Hussin and Ali (1987) describe two sites, where the concern again was the liquefaction potential of the 15 m or so of normally consolidated sands, silts and clays below a cemented crust about 2.5 m thick.

Trials were carried out using vibro-compaction and vibro-replacement, dynamic compaction and compaction grouting. The degree of improvement was assessed by three sets of tests: SPT, CPT and flat plate dilatometer. Hussin and Ali concluded that:

1. Vibro-replacement gave limited improvement in the silty sand below the cemented crust. It was better to use stone rather than sand as backfill. The method was more effective than dynamic compaction at depths greater than 8 m. There was no improvement in soils with more than 12 per cent fines or in the clayey soils.

2. Dynamic compaction with a 32 t weight and 30 m drop improved the silty sand below the cemented crust. Typical depths of improvement were about 10 m, but there was little or no improvement of cohesive soils, particularly those below the ground water table.

3. Compaction grouting was found to be effective in the sandy silts and silty sand layers. There was also some improvement in the silty sand below the cemented crust.

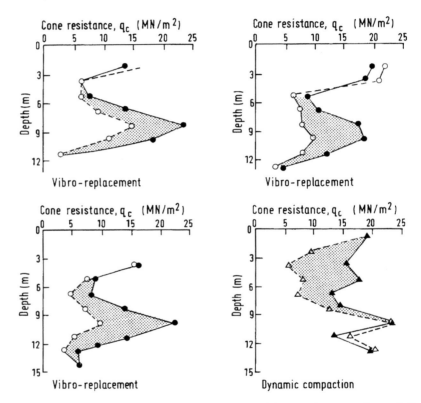

Figure A2.8 *Comparisons of mean cone resistances (averaged over 1.5 m depths) before (open points) and after (closed points) vibro-replacement (three sites) and dynamic compaction (one site) (after Keller et al, 1987)*

These few comparisons of techniques concentrate largely on the vibratory methods of compaction because these are the most widely used of the ground improvement methods. Further, this implies that the soils to be improved are granular and often hydraulically placed fills. Thus there is little published evidence of the effectiveness of different methods in treating heterogeneous fills (including wastes) or stratified natural deposits. It is again important to remember that some of the comparisons would not, perhaps, have been published had one or other technique proved less effective. In all cases there would have been inequalities in the basis of comparison; for example, was the cost per unit volume of treated ground the same? or was the *in-situ* test used as an indicator of improvement appropriate to the structural performance requirement of the treated soil/structure system?

For many reasons, therefore, it would be wrong to base a selection of method solely on brief descriptions given in this report. Rather, the lesson is that trials of different techniques or trials of one technique with variations of working procedures can be worthwhile. It is also apparent from the case history comparisons that the adopted method was not always the cheapest. Nor did the chosen method always produce the greatest improvement (in terms of the indicator testing) or the deepest improvement.

Table A2.4   *Summary of some dynamic compaction trials (after Keller et al, 1987)*

	Area A	Area B	Area C
Maximum energy per drop (kN-m)	2710	4070	7320
Total energy per unit area kN-m/m²	2920	6130	8760 (including 1170 from "ironing")
Imprint spacing and drop sequence	At 8 m centres, overlapping with three passes before "ironing"		
Surface preparation	0.6 m of silty sand placed on ground surface	0.45 m of silty sand formed above original ground surface during compaction	Upper 3 m of clayey sand removed and replaced with compacted silty sand
Groundwater	No dewatering: groundwater level close to ground surface	Groundwater lowered by about 1 m before compaction	Groundwater lowered by about 3 m before compaction
Densification in layer of poorly graded clayey sand 6–12 m	Objectives of densification not achieved at any depth	Objectives of densification met in upper portion only	Objective of densification met throughout the layer

A2.2   REFERENCES

Basore, C E and Boitano, J D (1969)
"Sand densification by piles and vibroflotation"
J Soil Mech and Found Div, Am Soc Civ Engrs, Vol 95, SM6, Nov, pp 1303–1323

Brown, R E and Glenn, A J (1976)
"Vibroflotation and Terra-probe comparison"
J Geotech Engg Div, Am Soc Civ Engrs, Vol 102, GT10, October, pp 1059–1072

Charles, J A, Earle, E W and Burford, D (1979)
"Treatment and subsequent performance of cohesive fill left by open cast ironstone mining at Snatchill experimental housing site, Corby"
Proc Conf on Clay Fills, London, 1978
Instn Civ Engrs, London, pp 63–72

Charles, J A, Burford, D and Watts, K S (1986)
"Improving the load carrying characteristics of uncompacted fills by preloading"
Municipal Engineer, Vol 3, No 1, pp 1–19

Choa, V, Vijiratnam, A, Karunaratne, G P, Ramaswamy, S D and Lee, S L (1979)
"Consolidation of Changi Marine Clay of Singapore using flexible drains"
Proc 7th Eur Conf Soil Mech and Found Engg, Brighton
British Geotech Soc, London, Vol 3, pp 29–36

Faraco, C (1981)
"Deep compaction field tests in Puerto de la Luz"
Proc 10th Int Conf Soil Mech and Found Engg, Stockholm, Vol 3, pp 659–662

Hussin, J D and Ali, S (1987)
"Soil improvement at the Trident Submarine facility"
In: *Soil improvement – a ten year update* (J P Welsh, ed)
Geotech Special Publication 12, Am Soc Civ Engrs, pp 215–231

Keller, T O, Castro, G and Rogers, J H (1987)
"Steel Creek dam foundation densification"
In: *Soil improvement – a ten year update* (J P Welsh, ed)
Geotech Special Publication 12, Am Soc Civ Engrs, pp 136–166

Mitchell, J K (1970)
"In-place treatment of foundation souls"
J Soil Mech and Founds Div, Am Soc Civ Engrs, Vol 96, SMI, January, pp 73–110

Mitchell, J K (1982)
"Soil improvement – state-of-the-art"
Proc 10th Int Conf Soil Mech and Found Engg, Stockholm, 1981
AA Balkema, Rotterdam, Vol 4, pp 509–565

Mitchell, J K (1986)
"Ground improvement evaluation by *in-situ* tests"
In: *Use of in-situ tests in geotechnical engineering*, Proc Spec Conf, Virginia
Geotech Special Publication 6, Am Soc Civ Engrs, pp 221–236

Mitchell, J K and Solymar, Z V (1984)
"Time-dependent strength gain in freshly deposited or densified sand"
J Geotech Engg Div, Am Soc Civ Engrs, Vol 110, No 11, November, pp 1559–1576

Murayama, S and Ichimoto, E (1985)
"Sand compaction pile (COMPOZER method for deep compaction)"
Proc Symp on recent developments in ground improvement techniques, Bangkok, 1982
AA Balkema, pp 79–84

Saito, A (1977)
"Characteristics of penetration resistance of a reclaimed sandy deposit and their changes
through vibratory compaction"
Soils and Foundations, Vol 17, No 4, Dect, pp 31–43

Schroeder, W L and Byington, M (1972)
"Experiences with compaction of hydraulic fills"
Proc 10th Ann. Symp on Engineering Geology and Soil Engineering, Moscow, Idaho,
pp 123–135

Solymar, Z V and Reed, D J (1986)
"A comparison of foundation compaction techniques"
Canad Geotech J, Vol 23, No 3, August, pp 271–280

Wallays, M (1985)
"Deep compaction by vertical and horizontal vibration"
In: *Recent developments in ground improvement techniques*
Proc Symp, Bangkok, 1982, AA Balkema, Rotterdam, pp 53–70